MARIE CURIE

THE PIONEER, THE NOBEL LAUREATE, THE DISCOVERER OF RADIOACTIVITY

DR RICHARD GUNDERMAN

WELBECK

THIS IS A WELBECK BOOK

Published in 2020 by Welbeck
an imprint of Welbeck Non-Fiction Limited,
part of the Welbeck Publishing Group
20 Mortimer Street
London W1T 3JW

A CIP catalogue for this book is available from the British Library.

ISBN 978-0-23300-617-8

Printed in Dubai

10 9 8 7 6 5 4 3 2 1

CONTENTS

PREFACE AND ACKNOWLEDGEMENTS

From discrimination and exclusion to the halls of power, from hidebound traditions to the dawn of a new era for women in education and science, from obscurity in Poland to one of the world's great centres of learning in Paris and on to worldwide fame, and from deep despair to love and then back again, Marie Curie's is the story of one of the greatest scientists the world has ever known, the greatest female scientist who ever lived, and by far the greatest dynasty in scientific history.

I have come to believe that the human mind, character, and imagination are shaped and invigorated less by lectures and exhortation than by example, and that stories – especially biographical narratives – hold great promise. We are unlikely to reach the heights of Marie Curie, but through her story we glimpse what it is like to work tirelessly in pursuit of a goal, to persevere in the face of grave losses, and above all to dream big.

For their patience and encouragement with my immersion in this project, I would like to thank my wife, Laura, and each of our children: Rebecca, Peter, David, and John. Numerous teachers and colleagues at Wabash College, the University of Chicago, and Indiana University have helped to spark and sustain this line of thought, as have many friends over the years. Marie Curie's story has helped me to understand that who we work with is at least as important as what we work on.

Pierre and Marie Curie in the laboratory.

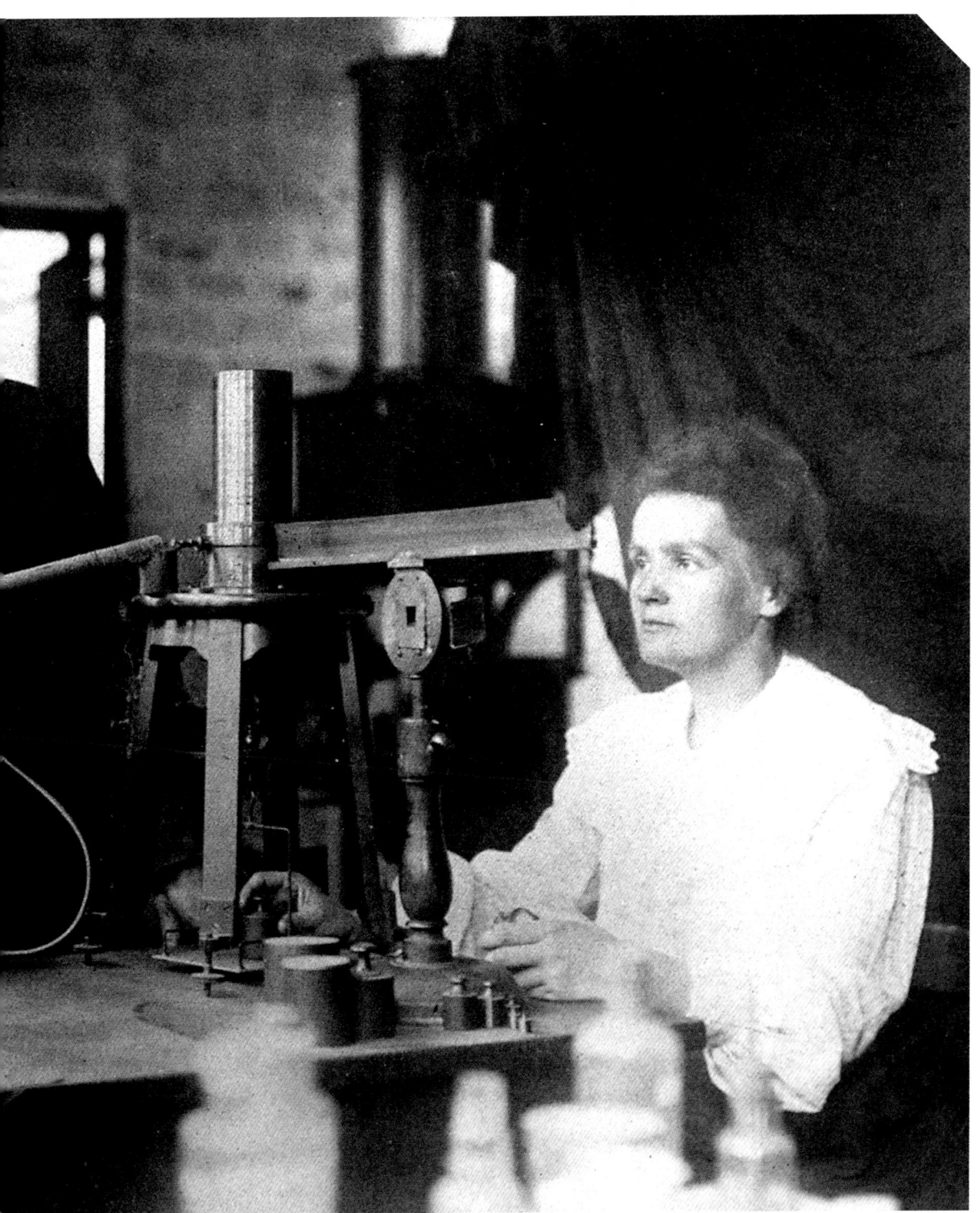

SCIENCE PRECEDING THE NINETEENTH CENTURY

THE CENTURY OF MARIE CURIE'S BIRTH, THE 1800s, WAS A MOMENTOUS ONE FOR SCIENCE. MOST NOTABLY PERHAPS, "SCIENCE" LARGELY REPLACED THE TERM FOR SUCH INVESTIGATIONS THAT HAD PREVAILED IN THE PREVIOUS CENTURIES, "NATURAL PHILOSOPHY", AND THE PEOPLE WHO PRACTISED IT CAME TO BE REFERRED TO AS "SCIENTISTS".

It is difficult to grasp the rapid change in scientific worldviews that took place during this time without first understanding some of the transformations in the preceding centuries.

In physics, Nicolaus Copernicus (1473–1543) demonstrated that the solar system is not earth-centred but sun-centred. Johannes Kepler (1571–1630) showed that the planets orbit the sun in ellipses. In addition to astronomical discoveries, Galileo (1564–1642) promoted experimentation as a central feature of the scientific method. Isaac Newton (1643–1727) introduced his laws of motion and gravitation, revealing as never before the power of mathematics as a tool for understanding the world.

In chemistry, Robert Boyle (1627–1691) helped alchemy – which aimed to turn base metals into precious ones, discover an elixir of immortality, and develop a universal solvent – give birth to chemistry, which sought to define the properties and structure of different forms of matter that we now call the chemical elements. Antoine Lavoisier (1743–1794), who died at the guillotine, founded modern chemistry and named the elements hydrogen and oxygen.

In biology and medicine, Andreas Vesalius (1514–1564) inaugurated a new era of anatomical study, encouraging learners to conduct their own dissections. William Harvey (1578–1657) demonstrated the circulation of the blood. The introduction of the microscope enabled Robert Hooke (1635–1703) to identify cells as a basic unit of living organisms. Later, both unicellular forms of life such as bacteria and multicellular forms such as human beings (each adult containing approximately 50 trillion cells) were identified.

Right: Sir Isaac Newton by Godfrey Kneller (1689).

SCIENCE IN THE NINETEENTH CENTURY

THE NINETEENTH CENTURY ITSELF WAS A TIME OF INCREDIBLE SCIENTIFIC FERMENT. IN MATHEMATICS, CARL GAUSS (1777–1855), ONE OF THE GREATEST MATHEMATICIANS OF ALL TIME, MADE SEMINAL CONTRIBUTIONS TO ALGEBRA, NUMBER THEORY, AND NON-EUCLIDIAN GEOMETRY, WHICH LATER PROVIDED AN IMPORTANT FOUNDATION FOR EINSTEIN'S THEORY OF RELATIVITY.

Above: James Maxwell.

George Boole (1815–1864) developed an approach to algebra and logic that required only the operators "and", "or", and "not", which later paved the way for computer science.

In physics, studies of charged particles by Michael Faraday (1791–1867) and James Maxwell (1831–1879) helped to create the field of electromagnetism, and Maxwell also made seminal contributions to the study of heat and its transfer, which came to be known as thermodynamics. Heinrich Hertz (1857–1894) detected electromagnetic radiation, leading eventually to radio and television, and he also discovered that when light strikes certain substances it causes electron emission, the photoelectric effect.

In chemistry, Dmitri Mendeleev (1834–1907), building on the atomic investigations of John Dalton (1766–1844), produced the periodic table of the elements. Friedrich Wöhler (1800–1882) and others founded organic chemistry, leading to the synthesis of hundreds of new compounds and introducing techniques that would play a crucial role in the study of the chemistry of living organisms, biochemistry. Among the compounds they synthesized were dyes, aspirin, and nitrogen-containing fertilizers.

Left: Dmitri Mendeleev in 1897.

Above: Thomas Edison c. 1922.

In biology, Charles Darwin (1809–1882) published his *On the Origin of Species*, arguing that species come and go depending on how well adapted they are to the environment in which they live, through the process of natural selection. John Snow's (1813–1858) studies of a cholera outbreak in London helped to advance the germ theory of disease, and Louis Pasteur (1822–1895) showed that vaccines could prevent the transmission of diseases.

Inventors devoted increasing attention to the technological possibilities of burgeoning science. Although Thomas Edison (1847–1941) did not invent the light bulb, he produced some of its most practical early examples, and he also helped to pioneer electric power generation, sound recording, and motion pictures. A rival of Edison, Nikola Tesla (1856–1943) produced working models of the AC (alternating current) induction motor, long-distance electrical transmission systems, and radio-controlled devices.

In addition to Edison and Tesla's work, inventors produced the first steam locomotive (1804), the first public railway (1825), the first electric motor (1829), telegraphy (1837), the introduction of the Bessemer process for the mass-production of steel (1859), the refining of oil (1856), and the invention of dynamite (1867), the typewriter (1870), the machine gun (1884), the automobile (1886), and the Swiss Army knife (1894).

WOMEN IN SCIENCE

It is notable that every scientist and inventor mentioned by name in this account so far is a man. Women tend to be absent from such accounts, for several reasons. First, men tended to be regarded as naturally better suited to scientific study, while women were expected to focus on domestic pursuits, such as the bearing and raising of children and the maintaining of a household. For many girls and young women, scientific work lay outside the ambit of their own imaginations.

Second, formal scientific education was largely unavailable to women. Only early in the 1800s were women admitted to learned societies, and it was not until later in the century that colleges and universities were founded in large numbers to educate women. This lack of educational opportunity prevented many otherwise-capable women from acquiring the basic knowledge and skills required for scientific work.

Third, the scientific contributions of women were often overlooked, a phenomenon that persisted well into the twentieth century. For example, physicist Joyce Bell Burnell discovered pulsars in 1967, but the Nobel Prize for the discovery went to her male supervisor. As Bell Burnell explained it, people simply assumed that only a senior man could be responsible for such discoveries, while women were supposed to function as "minions and junior staff, who weren't expected to think".

All this being said, however, by the time Marie Curie was born women had made many important scientific contributions. Hypatia of Alexandria (350–415) was an important figure in the history of mathematics and astronomy. Dorotea Bucca (1360–1436) held a chair in medicine at the University of Bologna for decades. Maria Merian (1647–1717) made foundational contributions to botany and entomology. And Maria Agnesi (1718–1799) was the first woman to hold a university chair in mathematics.

Above: Maria Agnesi.

MARIE'S POLISH HERITAGE

THE HISTORY OF POLAND IS AS INTEGRAL AS THE HISTORY OF SCIENCE IN UNDERSTANDING THE WORLD INTO WHICH MARIE CURIE WAS BORN. THE NATION'S FOUNDING IS USUALLY TRACED BACK TO 966, WHEN DUKE MIESZKO I ADOPTED CHRISTIANITY, WITH HIS SON BOLESŁAW I PROCLAIMING THE KINGDOM OF POLAND IN 1025.

Above: A contemporary map of Europe.

In 1569, Poland established a commonwealth with Lithuania, becoming an important European power notable for its parliament of nobles and its practice of electing kings.

Following an anti-Russian revolt, in 1772 the commonwealth underwent partition by its three neighbours, Prussia, Austria, and Russia. Reformers produced a constitution in 1791 that promised civil rights to the peasants, but Russia and Prussia sent in troops to prevent the liberalization, and Poland was subject to a second partition in 1793. An armed uprising in 1794 was crushed, leading the three countries to partition the country again. The nation of Poland disappeared from the map of Europe.

To garner Polish support for his campaign, in 1807 Napoleon created a Duchy of Warsaw, but the 1815 Congress of Vienna placed the kingdom under Russian rule. Resentment towards Russian rule led to an 1830 military revolt, which failed. A second revolt against Russian rule in 1863 was defeated, and much of Poland was annexed by Russia. Unable to secure independence, patriots refocused their efforts on promoting Polish culture through education and political activity.

At the time of Marie's birth in 1867, the people of Poland had been chafing for decades under what they regarded as foreign – and especially Russian – efforts to suppress their culture. The Russian and German languages became required for public communication and their cultures were promoted in the education of Polish children. The Catholic Church was suppressed. A non-violent wave of Polish

Above: Map of Europe in 1815.

nationalism arose, aiming to preserve and promote Polish identity.

The programme was termed "organic work". It sought to promote economic development by making Polish-owned businesses and farms more competitive. It also aimed to educate the Polish people by promoting literacy, establishing libraries, and disseminating publications. Instead of seeking to establish a new Polish nation through a popular uprising, proponents sought to enhance the culture's vitality in science, technology, and economics.

The Danish writer Georg Brandes visited Poland numerous times during Marie's life and provided rich descriptions of what life was like for the Polish people in the latter half of the nineteenth century. He described Poland as a nation that is "not only condemned to death, but which has been buried alive". Yet, he wrote, it "continually raises the lid of its coffin". The Polish people, in other words, sought to maintain their identity by every available means.

While travelling to Warsaw by train, Brandes was stopped and his belongings were searched by Russian officials. Many of his books were confiscated, and he received a receipt for "15 pounds of literature". In contrast to the "prudent and uniform" style of the Prussians, he described Russian rule as "incoherent, absurd, and often entrusted to clumsy hands". His reasoned arguments that he should be allowed to keep his books were met not with counterarguments but mere arbitrary authority.

Brandes described Warsaw as "next to Paris, once the most brilliant city in Europe", but lamented the fact that it was now little more than a "Russian provincial town". He accused the Russian government of seeking to humble

Right: Georg Brandes, by Peder Severin Kroyer (1900).

the pride of this capital of a country "the government does not wish to recognize" in every way possible. All vestiges of democracy had been erased, and "Russian Poland is altogether a country where nothing is elected."

"The Polish language is absolutely forbidden in the university," Brandes wrote, as well as in the streets and public offices. Signs were in Russian, and resistance to efforts to introduce the language into the churches led to the exile of several Polish bishops. He recounted the story of a 12-year-old boy who said to a schoolmate in Polish, "Let's go home together" and was punished by being held for 24 hours in the dark.

Finally, describing what is was like to learn to write under Russian censorship, Brandes wrote:

> *Almost all articles in which anything is really said are therefore not intended to be understood at the first reading. The language is abstract, vague, and of doubtful meaning. The whole public is taught to read between the lines. … I strove in vain to find expressions with double meanings; images, themselves indistinct, which could be understood by the audience; circumlocutions, which could be seen through and yet would be unassailable. … I became an expert in hints and implications.*

Both of Marie's parents came from families that had been part of the Polish nobility, whose lands and titles had been stripped from them after Poland's defeat at the hands of Russia. Her grandfather had promoted progressive views, arguing that the children of peasants should be educated right along with those of the nobility. One of her mother's brothers had been exiled to Siberia, and another had fled to France after being wounded twice in the war.

WŁADYSŁAW SKŁODOWSKI

Marie's father, Władysław Skłodowski, was the son of a teacher who followed in his father's footsteps, becoming a teacher of mathematics and physics and directing schools for boys. Russian suppression of Polish laboratory education led him to continue his teaching privately, using equipment in his home. His strong Polish patriotism eventually cost him his job, landing the family in difficult financial circumstances, and forcing them to take lodgers into their home.

Right: Władysław Skłodowski, father of Marie Curie, pictured with his three daughters, in 1890.

BRONISŁAWA SKŁODOWSKA

Marie's mother, Bronisława, had been educated at the only private school for girls in Warsaw. She became a teacher and later a headmistress but eventually relinquished her position due to her declining health. The Skłodowski children, even the girls, were expected to get an education and pursue careers that would enrich the lives of others, but Bronisława did not live to see what became of her youngest child. She died of tuberculosis when Marie was 10 years old.

Right: Bronisława Skłodowska, Marie's mother, 1860.

Russian suppression of Polish culture erected many barriers to the efforts of the patriotic Skłodowski family to advance themselves. Nevertheless, Marie's parents ensured that their children attended several different schools, where they received a rigorous education in languages, mathematics, and the sciences. Study of Polish language, culture, and history were illegal, but this did not prevent them ensuring that they were steeped in these subjects, both at school and at home.

BIRTH AND EDUCATION

BORN AS HER FAMILY'S FIFTH AND FINAL CHILD IN NOVEMBER 1867, MARIE WAS GIVEN THE NAME MARIA AND KNOWN TO HER FAMILY AS MANYA. IN CHILDHOOD, THE DEATHS OF MARIE'S OLDER SISTER AND MOTHER CAUSED GREAT SORROW BUT ALSO DREW THE FAMILY CLOSER TOGETHER.

Above: Joseph Skłodowski, Marie's brother.

As a student, Marie performed very well, receiving a gold medal on graduation from high school in 1883. In response to a bout of depression, her father ordered her to spend a year in the country, after which she recommenced her studies.

Marie's older brother Joseph was able to enrol in medical school at Warsaw, but she and her sister Bronya, as women, were ineligible. So they studied in the "floating university", so named because the time and location of its classes continually varied, to avoid detection by the Russian authorities. The students aimed to advance their education but also to contribute to a brighter future for Poland by building its intellectual and moral strength.

Recognizing the limitations of such studies, including the fact that they could never obtain a degree, Marie and Bronya devised a plan to advance their education. Bronya, the elder, would go to medical school in Paris, where women could be admitted for study, and Marie would remain in Poland and earn what money she could to help finance her sister's education. Later, once Bronya was earning a living, she would summon Marie to join her.

Right: Marie in 1883.

Above: Marie and her sister Bronya, 1883.

Marie started out tutoring the children of well-off families but soon realized that she needed to take a position as a governess. She went to work for the well-to-do Żorawski family outside of Warsaw, where she was also allowed to teach the children of the peasants. Soon after the oldest of the family's offspring, Kazimierz (1866–1953), returned home on break from his studies, he and Marie fell in love and decided to marry. His family hoped for a finer match, however, and forbade the union.

Despite her discomfort, Marie remained with the Żorawskis for two more years to earn money to send to Bronya. She described how she was managing to carry on in these terms:

> Creatures who feel as keenly as I do have to dissimulate as much as possible. I would give half my life to be independent again, to have my own home.

In fact, she appears not to have completely abandoned all hope of marrying Żorawski until she was 24 years old.

Marie remained in Poland for several years, doing what she could to advance her education. Describing the struggle to learn under such difficult circumstances, she wrote:

> I have a bright remembrance of the sympathetic intellectual and social companionship that I enjoyed at that time. Truly the means were poor and the results

less than considerable, yet I still believe that the ideas that inspired us then are the only way to genuine social progress. You cannot hope to build a better world without improving the individuals. To that end, each of us must work for his own improvement, and at the same time share a general responsibility for all humanity, as it is our duty to aid those to whom we can be most useful.

While Marie continued to work and study in Poland, her father obtained a position that paid sufficiently well to enable him to assume the financing of Bronya's education and repaying Marie for her contributions. Stimulated in part by her marriage to a Polish physician, Bronya reached out to Marie with an invitation to join her in Paris in 1890, but Marie felt obliged to decline, because she had not accumulated sufficient funds to cover the costs of her education. She wrote to her sister:

KAZIMIERZ ŻORAWSKI

In later years, Żorawski married a mathematician's daughter who was also an accomplished pianist, and he and his wife had three children. Professionally, he went on to earn a PhD in mathematics at Leipzig, became dean of the faculty at the university in Kraków, and eventually served as professor of mathematics in Warsaw. During the Nazi occupation of Poland, he was swept up in the Warsaw uprising and sent to a prison camp. After the war, he was elected to the Polish Academy of Sciences.

Right: Kazimierz Żorawski, 1888.

THE RADIUM INSTITUTE

Many years later, in 1932, the Radium Institute opened in Warsaw, directed by Marie's sister, Bronya. A statue of Marie was unveiled there in 1935, when the institute was renamed the Maria Skłodowska-Curie Institute of Oncology. It is said that in his later years, Żorawski, by now an eminent mathematician and "the first of his generation to bring the name of Poland to the forefront of mathematics", could often be seen sitting before the statue, perhaps imagining what might have been.

Below: Statue of Marie Curie in front of the Radium Institute (later Institute of Oncology) in Warsaw.

I have been stupid, I am stupid, and I shall remain stupid all the days of my life. I dreamed of Paris as a redemption, but the hope of going there left me a long time ago, and now that the possibility is offered to me I don't know what to do. I am exceptionally unhappy in this world.

Despite her bouts of deep discouragement, she wrote of what she had identified as a first principle: never to allow herself to be beaten down by persons or events. She would focus on "things, not persons", and she would remain true to her vision, even if circumstances seemed hostile to what she

aimed to accomplish. She would treat her lack of means and haphazard education not as insurmountable barriers but as wellsprings of determination upon which she could draw.

Finally, "realizing the dream that had been constantly in my mind for several years," Marie reached a point of readiness to commence her studies at the University of Paris. In the autumn of 1891, with barely sufficient funds secured, she set off. She had 40 roubles in her purse, a trunk, and a folding chair for the fourth-class train ride. By the time she took her seat in a classroom, eight years had passed since she had last been enrolled in a formal academic programme.

Below: The work room of the library in the Sorbonne.

UNIVERSITY OF PARIS

WHEN MARIE FIRST ARRIVED IN PARIS, SHE STAYED WITH BRONYA AND HER HUSBAND, BOTH OF WHOM WERE PHYSICIANS. ALTHOUGH THIS ARRANGEMENT HELPED HER SAVE MONEY, IT COST HER VALUABLE TIME, BECAUSE THE COMMUTE TO THE UNIVERSITY WAS APPROXIMATELY ONE HOUR IN EACH DIRECTION.

So Marie, as she had become known on her admissions application to the university, rented a small and inexpensive garret in the Latin Quarter, much closer to the university.

Decades later, Marie wrote of how she kept herself warm in the winter by piling all the clothes she owned on top of herself in bed. Once, she fainted in the library – an unmistakable indication, Bronya realized, that Marie was not eating sufficiently or obtaining adequate rest. As soon as she was nursed back to health at Bronya's, however, Marie returned to the Latin Quarter, in order to focus all her energies on her studies.

The University of Paris presented sharp contrasts to the educational environment Marie had known in Poland. In her home country, education had been highly regimented, and students faced relatively few choices. In Paris, she could attend any classes she wanted, sit or not sit for exams as she chose, and pay no fees except to take exams. The French system, she wrote later, aimed "to awaken the student's confidence in his own abilities and foster the habit of using them".

Marie described her life in a poem she wrote for a fellow Polish student, a portion of which reads:

Left: The Sorbonne, University of Paris.

How hard the life of her young years,
How rough her day till she retires
While, looking round, she sees her peers
Seeking new bliss with new desires.

Yet she has joy in what she knows
For in her lonely cell she finds
Richer air in which the spirt grows,
Inspired by the keenest minds.

Ideals flood this tiny room;
They led her to this foreign land;
They urge her to pursue the truth
And seek the light that's close at hand.

It is the light she longs to find,
When she delights in learning more.
Her world is learning; it defines
The destiny she's reaching for.

At the university, women were vastly outnumbered by men, and foreign women outnumbered those from France. This reflected the fact that secondary education in France was largely segregated, and the learning opportunities provided to girls were generally inferior to those available to boys. As a result, the total number of women studying at the University of Paris was approximately 200, out of a total student body of approximately 9,000.

The sex ratios were even more skewed in the sciences and mathematics. When Marie received her degree in science in 1893, just two years after her arrival in Paris, she was one of just two women. A year later, when she took her degree in mathematics, she was one of just five. These small numbers partly reflect the fact that a large proportion of the women at the university were merely attending classes but did not intend to pursue a degree.

This gender bias renders Marie's performance as a student even more remarkable. She knew well that she lacked the thorough science background of many of her fellow students, and she was taking her exams in a language that she had not yet mastered. When the exam results were read out, however, Marie finished first in her class, and when the results for the mathematics exams were posted the next year, Marie finished second.

During this time, Marie wrote to her brother about life as a student and what would be required to succeed in her investigations:

> *It is difficult for me to tell you about my life in detail; it is so monotonous and, in fact, so uninteresting. Nevertheless, I have no feeling of uniformity and I regret only one thing, which is that the days are so short and they pass by so quickly. One never notices what has been done; one can only see what remains to be done, and if one didn't like the work it would be very discouraging.*

> *It seems that life is not easy for any of us. But what of that? We must have perseverance and above all confidence in ourselves. We must believe that we are fitted for something, and that this thing, at whatever cost, must be attained. Perhaps everything will turn out very well, at the moment we least expect it.*

Marie's academic excellence and intellectual promise opened other doors for her. She received an Alexandrovitch scholarship that would fund more than a full year of study, and after that Gabriel Lippmann offered her a position in his laboratory studying the magnetic properties of various metal alloys.

MARIE'S TEACHERS

Marie had the opportunity to study with world-class mathematicians and scientists. One of her mathematics teachers was Henri Poincaré (1854–1912), perhaps the last mathematician to excel in all areas of the discipline. Another of her teachers was Joseph Boussinesq (1842–1929), who made important contributions in the physics and engineering of fluids. Poincaré was a theoretician, while Boussinesq was more of an experimentalist who emphasized practical demonstrations.

One of Marie's most important teachers was Gabriel Lippmann (1845–1921), who would receive a Nobel Prize for his contributions to colour photography five years after Marie received her first Nobel Prize. This is ironic, because Lippmann was one of four authors of a 1903 letter from the French Academy nominating Henri Becquerel and Pierre Curie for the Nobel Prize, but failing to mention Marie. It is possible that he thought that including a woman would doom the entire nomination.

Right: Henri Poincaré in 1912.

Later, using money she earned through a commission for a technical study, Marie repaid her Alexandrovitch stipend, something no one in the history of the scholarship had done up to that point.

Reflecting on these intense and impoverished years of her mother's life, Marie's daughter Ève later wrote, "She was proud of her poverty; proud of living alone and independent in a foreign city." Although these may not have been her happiest years, they were nonetheless "the most perfect in her eyes, the nearest to the summits of the human mission towards which her gaze had been trained". Ignoring her trials and privations, she was able to "magnify her sordid existence into magic".

Top: Joseph Boussinesq.

Above: Gabriel Lippmann in 1908.

Right: Marie Curie.

PIERRE CURIE

PIERRE CURIE'S FATHER, EUGENE, WAS A PHYSICIAN, AS WAS HIS FATHER. AS A YOUNG MAN, EUGENE HAD DREAMED OF DEVOTING HIS LIFE TO SCIENTIFIC WORK, BUT MARRIAGE AND THE BIRTH OF TWO SONS SOON BOUND HIM TO THE PRACTICE OF MEDICINE.

In the revolution of 1848, during which he suffered a fractured jaw, he received a medal for his service. Later he accepted a position with a child welfare organization, which allowed his boys to grow up in the suburbs.

Pierre's mother, Claire, came from a prominent manufacturing family that had invented several processes for making dyes. However, the family suffered financial ruin during the 1848 revolution. Decades later, Marie wrote that her mother-in-law had "accepted with tranquil courage the precarious conditions which life brought her and gave proof of an extreme devotion as she made life easier for her husband and children by her activity and good will".

Pierre was born in May of 1859, three and a half years after the birth of his older brother, Jacques (1855–1941). Pierre was entirely home-schooled, in part because, as Marie described it, "his intellectual capacities were not those which would permit the rapid assimilation of a prescribed course of studies." The young Pierre thought himself "slow", but in fact his was the sort of mind that, instead of learning many subjects at once, needed to concentrate intensely on a single problem.

At age 14, Pierre was assigned a tutor who taught him mathematics, and he soon earned a bachelor of science degree at the young age of 16. He began attending lectures on physics at the University of Paris, intending to take an advanced degree. Financial circumstances prevented this, however, and instead he began work as the supervisor of the physics laboratory, overseeing the exercises of undergraduate students. He remained in this position for five years.

While both Pierre and Jacques were working in the sciences at the university, they enjoyed taking long walks in the Parisienne countryside. There Pierre could follow his train of thought without distraction, an opportunity that frequently eluded him in Paris. He wrote in his diary,

> When, in the process of turning slowly upon myself, I try to gain momentum, a nothing, a word, a story, a paper, a visit stops me and is able to put off or retard forever the moment when, granted a sufficient swiftness I might have, in spite of my surroundings, concentrated my own intention. We must eat, drink, sleep, be idle, love, touch the sweetest things of life and yet not succumb to them. It is necessary that, in doing all this, the higher thoughts to which one is dedicated remain dominant and continue their unmoved course in our poor heads. It is necessary to make a dream of life, and to make of a dream a reality.

Soon Pierre and Jacques, who had completed his advanced degree, began investigating crystals, which led them to the discovery in 1880 of piezoelectricity. *Piezo* is the Greek

PIEZOELECTRICITY

In a nutshell, the phenomenon of piezoelectricity reflects the fact that crystals are composed of regular, repeating arrangements of atoms, with balanced positive and negative charges. When pressure is applied, the structure deforms, upsetting the balance and causing net charges to appear. In reverse, the application of an electric potential will cause the atoms to rearrange themselves slightly, causing a deformation in shape.

The piezoelectric effect can be used to make transducers, devices that convert one form of energy into another. Of course, not all transducers are products of human invention. The human body contains many transducers, such as the retina of the eye, which converts light energy into electrical impulses in the optic nerve, and the inner ear, which enables sound vibrations to generate electrical signals in the eighth cranial nerve.

Today many inventions rely on the piezoelectric effect. A microphone converts sound energy to electric signals, and a speaker converts electrical impulses to sound. Quartz clocks rely on the fact that quartz crystals subjected to an electric potential oscillate at a characteristic frequency, making it possible to measure the passage of time with a high degree of accuracy. The piezoelectric properties of zinc oxide account for its wide use in semiconductors, the backbone of digital electronic circuits.

Left: Piezoelectric balance presented by Pierre Curie to Lord Kelvin.

word for pressure, and the Curies showed that when a crystal such as quartz is deformed, it generates an electric potential. Lippman predicted that the converse would also be the case, and the Curies soon achieved experimental proof. From this they produced a quartz piezoelectric electrometer, crucial later in studying radioactivity.

In 1883, the Curie brothers were split apart, Jacques moving to Montpelier in the south of France to become the head lecturer in minerology and Pierre becoming the director of the laboratory in the School of Industrial Physics and Chemistry in Paris. Though separated by space, however, the brothers remained bound together in spirit. Of their frequent reunions, Pierre wrote:

At times it seemed to me that we had gone back to the days when we lived entirely together. Then we always arrived at the same opinions about things, with the result that it was no longer necessary for us to speak in order to understand each other.

As director of the laboratory, Pierre was little older than many of his students, and he later recalled with amusement how one day, staying late in the lab, he and the students found the door locked. To get out, they had to climb down to the first floor using a pipe that ran along one of the windows. Directing the work of several dozen students occupied nearly all Pierre's energies, yet he found time to continue the study of crystals, publishing several important papers.

Pierre finally produced his doctoral thesis, "The Magnetic Properties of Bodies at Various Temperatures", in 1895, under the supervision of Lippmann. Having impressed important physicists such as Lord Kelvin, a new chair was created for him at the School of Industrial

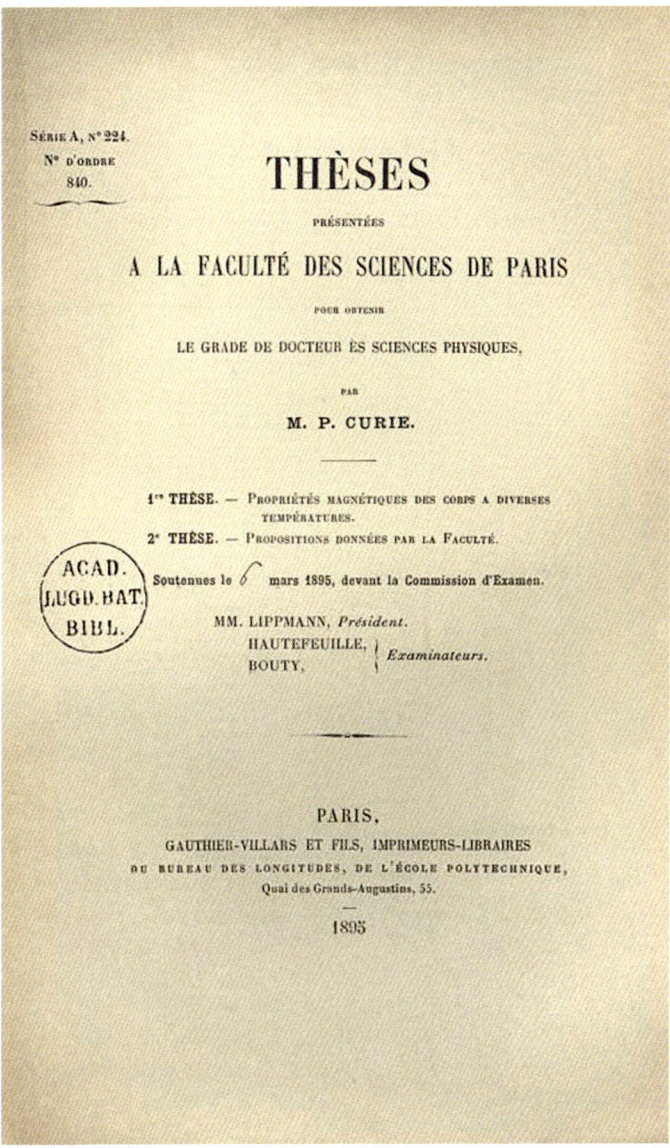

Physics and Chemistry. Despite attaining the title of professor, however, little was done to improve the quality of Pierre's research facilities, and he soldiered on under the same difficult conditions.

Above: Cover of Pierre Curie's 1895 doctoral thesis.

COURTSHIP AND MARRIAGE

IT IS NOT SURPRISING THAT PIERRE, THOUGH IN HIS MID-30S, HAD NOT MARRIED. MANY OPPORTUNITIES TO MEET AND COURT YOUNG WOMEN WOULD HAVE PRESENTED THEMSELVES, BUT HE WAS DEDICATED TO HIS SCIENCE.

What others might have regarded as inducements to marriage did not sway him so strongly. He seems to have regarded romance as a temptation to be resisted, out of dedication to his scientific investigations.

At 22, he had written in his diary:

> Woman loves life for the living of it far more than we do: women of genius are rare. Thus, when we, driven by some mystic love, wish to enter upon some anti-natural path, when we give all our thoughts to some work which estranges us from the humanity nearest us, we have to struggle against women. The mother wants the love of her child above all things, even if it should make an imbecile of him. The mistress also wishes to possess her lover and would find it quite natural to sacrifice the rarest genius in the world for an hour of love. The struggle almost always is unequal, for women have the good side of it: it is in the name of life and nature that they try to bring us back.

Marie and Pierre were introduced in 1894 by a mutual acquaintance, Joseph Wierusz-Kowalski (1866–1927), a Polish physicist who was lecturing in Paris, accompanied by his wife. During a conversation, Marie told him of the difficulty she was experiencing with her research on the magnetic properties of metals, largely due to the cramped conditions in Lippmann's laboratory. The physicist told her about a young scientist he knew who might

be able to provide more suitable lab space, or at least offer some helpful advice.

Marie and Pierre met the next evening in the physicist's rooms for tea. Marie described the moment she first laid eyes on her future husband:

> As I entered, Pierre Curie was standing in the recess of a French window opening on a balcony. He seemed to me very young, though he was at that time 35 years old. I was struck by the open expression of his face and the slight suggestion of detachment in his whole attitude.
>
> His speech, rather slow and deliberate, his simplicity, and his smile, at once grave and youthful, inspired confidence.
>
> We began a conversation, which soon became friendly. It first concerned scientific matters about which I was very glad to be able to ask his opinion. Then we discussed certain social and humanitarian subjects which interested us both. There was, between his conception and mine, despite the difference between our native countries, surprising kinship, no doubt attributable to a certain likeness in the moral atmosphere in which we were both raised by our families.

Right: Pierre and Marie Curie in 1895.

Meeting again at the Physics Society and in the laboratory, Pierre asked if he might call on her. Soon he began speaking to Marie of his dream of "an existence consecrated entirely to scientific research", asking her to share that life. To pursue such a dream would require Marie to abandon her plans to return to Poland and her father at the conclusion of her studies. Visiting Poland on vacation, she maintained a correspondence with Pierre that strengthened the bond between them.

He wrote to her:

It would be a beautiful thing in which I hardly dare to believe, to pass through life together hypnotized by our dreams: your dream for your country; our dream for humanity; our dream for science. Of all these dreams, I believe the last alone is legitimate. … I strongly advise you to return to Paris in October. I shall be very unhappy if you do not come this year. … I ask it because I believe you will work better here and that you can accomplish here something more substantial and more useful.

Many might find such a letter disappointing, but not Marie. As she explained it,

For Pierre Curie there was only one way of looking at the future. He had dedicated his life to his dream of science: he felt the need of a companion who could live his dream with him. He told me many times that the reason he had not married until he was 36 was because he did not believe in the possibility of a marriage which would meet this, his absolute necessity.

After Marie returned to Paris, each realized that it would be impossible to find a better life companion. They decided to get married, and the ceremony took place in July 1895. It was, according to Marie, "the simplest ceremony possible", and a civil one, since Pierre professed no religion and she did not practise any. In attendance were Pierre's parents, who received their new daughter-in-law with "great cordiality", and Marie's father and sisters.

Upon their return to Paris, the newlyweds moved into a three-bedroom apartment not far from the School of Physics. Its chief attraction was a view of a large garden. They furnished it with donations from their families. Lacking resources, Marie assumed "practically all the household duties", a situation familiar to her from her student days. Both were happy to economize, and their greatest challenge was to cram all their daily work into the short space of 24 hours.

Marie approached her responsibilities as a new wife as she did everything else. She arranged to take cooking lessons. She practised preparing meals, even though the absent-minded Pierre sometimes seemed not to notice. She read and re-read her cookbooks, carefully noting in their margins her failures and successes. Even in the kitchen, she was experimenting.

The couple spent virtually all their time together, and as a result they produced very few lines of correspondence. On rest days and vacations, they would walk or bicycle in the countryside or the mountains. However, Pierre found it difficult to remain for long anywhere that lacked the facilities to continue his work. Within a few days, he would say, "It seems to me a very long time since we accomplished anything."

HONEYMOON

The couple honeymooned with a cross-country bicycle tour, a plan made feasible by the recent introduction of the inflatable inner tube. Though their means were modest, as their daughter Ève would later describe it, "At the cost of some thousands of pedal strokes and a few francs for village lodgings, the young couple attained the luxury of solitude shared between them for long enchanted days and nights." During their excursions they talked of sharing the same laboratory. They would be together always.

Below: Pierre and Marie Curie on their honeymoon.

THE INVISIBLE LIGHT

FOR MOST OF HUMAN HISTORY, VISIBLE LIGHT WAS THE ONLY KNOWN FORM OF WHAT WE NOW CALL ELECTROMAGNETIC RADIATION. THE ANCIENT GREEKS KNEW THAT IT TRAVELLED IN STRAIGHT LINES AND COULD BE BOTH REFLECTED AND REFRACTED.

It was not until 1800 that new portions of the electromagnetic spectrum began to be discovered, with William Herschel's (1738–1822) detection of infrared radiation. A year later, Johann Ritter (1776–1810) discovered ultraviolet radiation.

It was in 1845 that Michael Faraday connected these forms of radiation to electromagnetism, showing that an electromagnetic field could affect the polarization of light. In the 1860s, James Maxwell developed equations to describe electromagnetic fields and realized that such waves travel at the speed of light, leading him to hypothesize that light itself is a type of electromagnetic wave. By 1886, Heinrich Hertz was both producing and detecting radio waves.

In November of 1895, the same year that Marie and Pierre were married, a German physicist named Wilhem Roentgen (1845–1923) was performing experiments that involved passing high-voltage currents through evacuated tubes. What he discovered would revolutionize not only physics but also fields such as chemistry, medicine, and astronomy, and Roentgen himself would receive the very first Nobel Prize in Physics in 1901 "for the discovery of the remarkable rays subsequently named after him".

The son of a German father and a Dutch mother, Roentgen was both colour blind and blind in one eye as a result of a childhood accident. Expelled from high school for a

Left: Portrait of Michael Faraday by Thomas Phillips (1843).

Above: Heinrich Hertz.

WILHELM ROENTGEN

Other investigators, such as Philipp Lenard and Nikola Tesla, had probably observed x-ray phenomena before, but it was Roentgen who recognized the significance of what he was seeing and set about systematically investigating the new rays. The discovery came on 8 November 1895, when he noticed that cathode rays produced fluorescence on a screen painted with barium salts. Intrigued, he practically lived in his laboratory for the next six weeks, conducting various experiments.

Roentgen named the new rays "x-rays", based on the use of the letter x to denote the unknown. About two weeks into his investigations, he created an x-ray image of Bertha's ringed hand, the first x-ray of a human being ever produced. It is said that when she saw the skeleton-like image, she remarked, "I have seen my death." Roentgen's first paper on x-rays was published in 1895, and news of the discovery spread like wildfire around the world.

Roentgen declined to attempt to patent the process by which he produced x-rays, and he donated his Nobel Prize stipend, over $1 million in today's terms, to his university. Just months after his discovery, Bertha wrote, "It is not easy to be a famous man, and very few people have any conception of how much work and excitement it carries with it. ... [But you know] that his greatest recompense lies in the fact that he could find something valuable in the field of pure science."

Roentgen went on to assume the physics chair at the University of Munich. He expected to emigrate to the United States, but his plans were dashed by the outbreak of World War I. He ended up spending the rest of his life in Germany. His savings were consumed by post-war inflation, and he lived out his last years in poverty. He died in 1923 from colon cancer. After his death, an element, roentgenium, was named after him, and the roentgen is now an international unit of x-ray exposure.

Above: Wilhelm Roentgen, 1899.

prank, he had difficulty gaining admission to university, but eventually earned his PhD from the University of Zurich. He and his wife, Bertha, adopted a daughter. The following year, 1888, he was awarded the chair of physics at the University of Würzburg, where he made his great discovery.

Roentgen foresaw the use of x-rays in medicine, where the ability to peer inside the living human body would revolutionize the diagnosis of injuries and diseases. Decades later, around 1970, a British engineer named Godfrey Hounsfield (1919–2004) would devise the CT (computed tomography) scanner, which uses x-ray beams passed through the body from many different directions to produce far more precise images of internal anatomy, for which he received a Nobel Prize in 1979.

In 1912, physicists realized that while the wavelengths of visible light were too great to permit the study of the molecular structure of crystals, shorter-wavelength x-rays could work very well. A father and son team, William Henry Bragg (1862–1942) and William Lawrence Bragg (1890–1971), developed Bragg's Law of x-ray diffraction for determining crystalline structure. The Braggs received the 1915 Nobel Prize in Physics.

During the 1950s, x-ray crystallography played a crucial role in determining the structure of DNA, the principal conveyor of genetic information. Using x-ray crystallographic data produced by Maurice Wilkins (1916–2004) and Rosalind Franklin

Right: Godfrey Hounsfield, 1975.

(1920–1958), James Watson (b. 1928) and Francis Crick (1916–2004) were able to show that DNA, the carrier of genetic information, has a double-helix structure. Watson, Crick, and Wilkins shared the Nobel Prize in Physiology or Medicine in 1962.

Another crucial role for x-rays emerged in astronomy. X-ray astronomy only became possible around 1960, when x-ray detectors could be positioned above the earth's atmosphere. While celestial bodies such as the sun emit a substantial amount of energy in the visible portion of the electromagnetic spectrum, much more powerful bodies such as neutron stars and black holes produce emissions in the x-ray and gamma-ray portions of the spectrum.

Today we know that all the hues we can see with our eyes make up only about 0.0035 per cent of the electromagnetic spectrum, which ranges over frequencies that vary by at least 27 orders of magnitude. Extremely low-frequency radio waves have frequencies of several Hertz and wavelengths of approximately 100,000 km, while at the opposite end of the spectrum gamma rays can have frequencies up to 4×10^{27} Hz and wavelengths shorter than a picometre (one-trillionth of a metre).

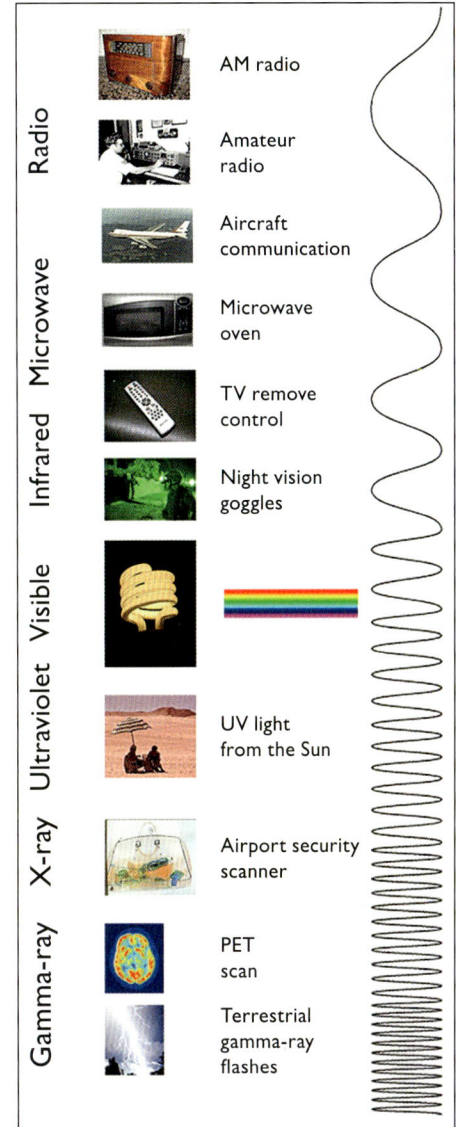

Radio	AM radio
	Amateur radio
	Aircraft communication
Microwave	Microwave oven
	TV remove control
Infrared	Night vision goggles
Visible	
Ultraviolet	UV light from the Sun
X-ray	Airport security scanner
Gamma-ray	PET scan
	Terrestrial gamma-ray flashes

Left: James Watson and Francis Crick with a model of the double-helix structure of DNA, 1953.

Above: The electromagnetic spectrum.

BECQUEREL

HENRI BECQUEREL (1852–1908) MAY NOT HAVE BEEN THE FIRST TO OBSERVE THE PHENOMENON LATER KNOWN AS RADIOACTIVITY, BUT HIS WORK INSPIRED THE INVESTIGATIONS OF MARIE AND PIERRE CURIE. HAD BECQUEREL NOT BEEN AT WORK ON IT, IT IS UNLIKELY THAT THE CURIES WOULD EVER HAVE TAKEN IT UP.

A perfect balance was needed – Becquerel had to be interested enough in the phenomenon to note some puzzling features, but not so intrigued that he wished to continue the investigation himself. Henri Becquerel was a child of privilege, not only scientifically but also economically. He was educated in the best schools. His first wife died giving birth to their son, Jean, who would later become the fourth consecutive Becquerel to serve as professor of physics at the Museum of Natural History. One of Henri's greatest interests was the study of phosphorescence, the relatively slow release of light by certain materials when they are exposed to light, perhaps best known through "glow-in-the-dark" toys.

When Becquerel learned of Roentgen's discovery of the x-ray, he began investigating the possibility that it was somehow connected with the phosphorescence on which he had already been working. Perhaps certain materials, such as uranium salts, would emit not only visible light but x-rays when they were "charged" by an energy source such as sunlight. To test his hypothesis, he exposed uranium salts to sunlight, then placed them on a shielded photographic plate, which they exposed.

This suggested that the uranium was indeed emitting a penetrating form of radiation of the sort Roentgen had described. During a period of poor weather, however, when he was unable to expose the uranium salts to sunlight, Becquerel was surprised to discover that the shielded photographic plate was nevertheless exposed. This suggested that the radiation emitted from the uranium salts was not stored energy from sunlight, but energy intrinsic to the uranium itself.

Although Becquerel published seven papers on the new phenomenon in 1896 and two more in 1897, by 1898 he seems to have lost interest in the subject, in part because x-rays produced by cathode-ray tubes of the sort used by Roentgen seemed to produce a superior version of the same effect and in a much shorter period of time. Though Becquerel had laid an important foundation for the understanding of radioactivity, he would allow a graduate student, Marie Curie, to take up the problem.

Later, when Pierre and Marie isolated radium, Becquerel kept a small quantity of the new element in a test tube in his vest pocket. To his surprise, he developed burns on his skin just below where he kept

Right: Henri Becquerel.

A FAMILY OF SCIENTISTS

Henri Becquerel was born into a family of important scientists. His grandfather, Antoine César Becquerel (1788–1878), was a pioneer in research involving electricity, developing the first technique for extracting elements from ores using electrolysis and producing important studies on luminescence. He became professor of physics at the French Museum of Natural History, was elected a fellow of the Royal Society, and his is one of 72 names inscribed on the Eiffel Tower.

Above: Edmond Becquerel.

Becquerel's father, Edmond (1820–1891), collaborated with his grandfather in producing the photovoltaic effect, the conversion of light energy into electricity, which is sometimes known as the Becquerel effect. This is the underlying principle of what is commonly known as solar energy. Like Antoine, Edmond became professor of physics at the Museum of Natural History, and his treatise, "Light, Its Causes and Effects", became the standard text for many years.

Left: Antoine Becquerel.

the radium. When Pierre learned of this, he intentionally exposed his arm to the element for several hours, developing a similar burn that required months to heal. These initial observations led later to the use of radioisotopes in the treatment of diseases such as cancer.

Although Becquerel largely abandoned the study of radioactivity, for which he would later share a Nobel Prize with Pierre and Marie Curie, he continued to make notable scientific contributions. For example, he demonstrated that beta particles have the same charge-to-mass ratio as electrons. He also showed that, over time, uranium loses its radioactivity, which led to the discovery that radioisotopes undergo decay into non-radioactive forms.

Another major inducement to the Curies' investigations came from the Irish physicist William Thomson (1824–1907), better known as Lord Kelvin. While Becquerel had focused on uranium's effects on photographic plates, Kelvin studied uranium's ability to electrify air. When Marie later set out on her doctoral research, she was not attempting to identify a new form of radiation, but to measure as precisely as possible the magnitude of this effect.

Of course, Marie was not the only person to study such phenomena. For example, the New Zealand physicist Ernest Rutherford (1871–1937), later known as the father of nuclear physics, was hard at work on the phenomenon of radioactivity. Working at McGill University in Canada, Rutherford's "investigations into the disintegration of the elements and the chemistry of radioactive substances" would garner him the 1908 Nobel Prize in Chemistry.

Describing the sense of urgency with which he and his colleagues worked, Rutherford later wrote to his mother in 1902:

Above: Lord Kelvin.

Above: One of Becquerel's photographic plates, fogged by exposure to uranium.

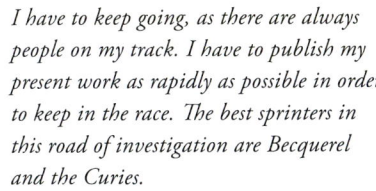

I have to keep going, as there are always people on my track. I have to publish my present work as rapidly as possible in order to keep in the race. The best sprinters in this road of investigation are Becquerel and the Curies.

It was in fact after he received the Nobel Prize that Rutherford made his most important scientific contribution, when in 1911 he presented what became known as the Rutherford model of the atom, proposing that electrons orbit the nucleus. Marie Curie differed considerably from Rutherford in this respect, being much less inclined to engage in theoretical speculation and model building and more focused on producing practical experimental results.

Left: Ernest Rutherford.

BIRTH OF IRÈNE

DURING THE EARLY YEARS OF THE CURIES' MARRIAGE, MARIE FULFILLED THE ROLES OF HOMEMAKER AND SCIENTIST.

In the days before refrigeration, it was necessary to go to the market every day, a duty she attended to first thing in the morning. She prepared meals and kept house, a responsibility in which she enjoyed one great advantage: she and Pierre were relatively indifferent to furnishings and the like, which left her with a simpler household to maintain.

Decades later, Marie's daughter Ève reported that one of Marie's foremost incentives as a homemaker was the mortification she would experience if her mother-in-law ever found reason to question her cooking or housekeeping. On some occasions Marie prepared very simple dishes to save time, and at other times she left dishes cooking on low heat throughout the day. The two or three hours she devoted each day to homemaking left her with about eight hours for her research.

In the second year of their marriage, 1897, Marie became pregnant. She longed for a child, but pregnancy soon imposed considerable limits on her ability to work. In March of that year, she wrote to one of her sisters:

> *I am very late with my birthday letter, but I have been very unwell all these last weeks, and that deprived me of the energy and freedom of mind for writing. I am going to have a child, and this hope has a cruel way of showing itself. For more than two months I have had continual dizziness, all day long from morning to night. I tire myself out and get steadily weaker, and although I do not look ill I feel unable to work and am in a very bad state of spirits. My condition irks me now because my mother-in-law is now seriously ill.*

Pierre's mother had been diagnosed with advanced breast cancer. Weighing on Pierre's and particularly Marie's mind was the possibility that her illness would spoil everyone's spirts, especially if the birth coincided with her death. When summer came, it was decided that Marie would leave Paris and go to the coast. Pierre remained behind to finish up his teaching duties and help his father care for his ailing mother.

Marie's daughter Irène was born in September. Dr Curie, the father-in-law, attended the delivery, during which Marie was said to have barely uttered a sound. Due to her mother-in-law's illness, the family was largely unavailable to assist in the latter months of the pregnancy. After the baby's birth, Marie was forced to hire extra help, which drove up expenses. Just two weeks after the baby's birth, Marie's mother-in-law died.

Right: Pierre, Irène, and Marie Curie, c. 1902.

Marie faced a range of challenges: balancing her career, her marriage, and her new-found role of mother. Writing to her sister-in-law, Marie expressed her anxieties:

I am still nursing my little Queen, but lately we have been seriously afraid that I could not continue. For three weeks, the child's weight has suddenly gone down, Irène looked ill, and was depressed and lifeless. For some days now, things have been going better. If the child gains weight normally I shall continue to nurse her. If not, I shall take a nurse, in spite of the grief this would be to me, and in spite of the expense; I don't want to interfere with my child's development for anything on earth.

Despite Marie's hopes to the contrary, she was soon forced to give up nursing the baby, and a wet nurse was hired. True to form, Marie began keeping a notebook, tracking all the notable events and milestones in her daughter's life, a practice she would later repeat to a lesser degree after Ève's birth. She tracked Irène's weight, noted when she began to move about on her own, how she reacted to strangers, and eventually, the eruption of her first tooth.

Years later, Marie revealed that she did not even think of abandoning her scientific work. She wrote:

Such a renunciation would have been very painful to me, and my husband would not even think of it; he used to say that he had gotten a wife made expressly for him to share all his preoccupations. Of course, we had to have a servant, but I personally saw to all the details of the child's care. While I was in the laboratory, she was in the care of her grandfather, who loved her tenderly and whose own life was made brighter by her. So the close union of our family enabled

me to meet my obligations. Things were particularly difficult only in case of more exceptional events, such as a child's illness, when sleepless nights interrupted the normal course of life.

Fortunately, Pierre's father, now a widower, had proved eager to help raise the baby. He soon came to live with them, bearing the lion's share of the responsibility for child rearing. This provided Marie with time to continue her investigations, including the publication of her first paper, on the magnetic properties of metals. Now she could turn her attention to another priority: obtaining her doctorate. All that was needed was a topic for her research.

Above: Pierre, Irène, and Marie Curie.

Opposite: Eugene Curie, early 1900s.

RADIUM AND POLONIUM

BECQUEREL HAD SHOWN THAT, MUCH LIKE VISIBLE LIGHT, EMISSIONS FROM URANIUM COULD EXPOSE A PHOTOGRAPHIC PLATE. HE ALSO DEMONSTRATED THAT URANIUM RETAINS THIS ABILITY FOR AT LEAST A PERIOD OF SEVERAL MONTHS.

Left: Curie quartz piezoelectrometer.

We now know that one isotope of the element, uranium 235, has a half-life of about 700 million years, while that of uranium 238, which now makes up more than 99 per cent of the uranium on earth, has a half-life of about 4.5 billion years.

Marie and Pierre saw this new phenomenon as an opportunity to explore an "entirely new" phenomenon on which nothing had been written. But where could they carry out their investigations? Pierre obtained permission from the director of his school to use a glass-walled study on the ground floor, then being used as a storeroom. To measure the ionization of air by uranium, the Curies used an apparatus that combined a Curie electrometer, a piezoelectric quartz crystal, and an ionization chamber.

Recording her observations in a notebook supplied by Pierre, Marie established that she could precisely measure the degree of radiation produced by uranium, and that the phenomenon is unaffected when uranium is combined with other elements or exposed to changes in light and temperature. Testing a variety of additional elements, she determined that thorium was the only other known element that emitted similar rays. To describe this property, she coined the term "radioactivity".

PITCHBLENDE

They decided to focus their efforts on pitchblende, a uranium ore now known as uranite, which seemed to be at least four times more radioactive than its uranium content could account for. Mined in locations such as the border between Germany and Czechoslovakia, the "pitch" in pitchblende denotes its black colour. Uranium itself is a dense, grey metal that has the highest atomic weight of any of the elements whose existence predates the formation of the earth.

Their method was elegant but entailed a great deal of painstaking effort. They would break down pitchblende into its various chemical components and then test the radioactivity of each one. With each successive separation, the radioactive element should become more and more concentrated and therefore more radioactive. As they proceeded, they realized that two elements were involved, one bound to bismuth and the other to barium.

Each element seemed to be hundreds of times more radioactive than uranium, and they knew that they needed to isolate each in pure form. Because the concentrations of the elements were so low, they would need much larger quantities of pitchblende than were available to them at the time. They would also need a much larger space in which to work, preferably a well-ventilated one, since heating the pitchblende gave off noxious gases.

The space problem was solved when Pierre secured a large but poorly furnished hangar that had been used for cadaver dissections by medical students. Sufficient quantities of pitchblende were secured when tons of the substance, a by-product of silver mining, were obtained from Austria. Considered worthless, it had been left to leech into the soil of a pine forest. To the Curies, the sacks of "brown dust mixed with pine needles" represented a godsend.

To their surprise, the Curies found that other minerals were radioactive, and that even ones that contained uranium exhibited a much higher than expected degree of radioactivity. Marie hypothesized that these other minerals contained minute amounts of one or more as-yet-undiscovered radioactive elements. Pierre, intrigued by this suggestion, set aside his work on crystals and began collaborating with Marie to identify this unknown element.

The Curies divided their labours. Pierre would focus on describing the properties of the new elements, and Marie would continue her efforts to prepare them in pure form. It would later become clear that radium is present in pitchblende at a concentration of about one-seventh of a gram per ton. Their efforts were like trying to find a needle in a haystack, except that the elements they sought were actually mixed in throughout the ore they were handling. Describing her work, Marie wrote:

I had to work with as many as 20 kilograms of the material at a time, so that the hangar was filled was great vessels full of precipitates and liquids. It was

exhausting work to move the containers about, to transfer the liquids, and to stir for hours at a time, with an iron bar, the boiling material in the cast-iron basin.

The Curies invented names to describe the elements they were chasing. The first they called polonium, after Marie's motherland. The second they called radium. As Marie's work continued, it became clear that the radium would be easier to isolate than the polonium, so they concentrated their efforts on it. A special source of joy in the work was the discovery that concentrated radium was luminous – that is, it constantly emits a faint, bluish glow, even without being heated.

In fact, what the Curies had isolated was not pure radium, but radium chloride. Radium was not isolated in its pure metallic state until some years later, in 1911, by André-Louis Debierne (1874–1949) and Marie, using the technique of electrolysis. Completing the work of isolating radium chloride in 1898, in the following years the Curies published papers describing their work, which they also presented at the International Congress of Physics, held in Paris in 1900.

Although the work was very hard, Marie seemed to be as happy as she would be at any point in her life. She wrote:

A great tranquillity reigned in our poor, shabby hangar; occasionally, while observing an operation, we would walk up and down talking of our work, present and future. When we were cold, a cup of hot tea, drunk beside the stove, cheered us. We lived in a preoccupation as complete as that of a dream.

Left: Radium watch hands under ultra-violet light.

Opposite: Pierre and Marie Curie in their laboratory.

DIFFICULT YEARS

IN THE PERIOD SPANNING THE LAST FEW YEARS OF THE NINETEENTH CENTURY AND THE FIRST FEW OF THE TWENTIETH, THE CURIES FACED A NUMBER OF HARDSHIPS.

Money presented one challenge. Pierre's meagre salary barely covered their expenses, and after the birth of Irène they were spared financial shortfalls only by the receipt of occasional prizes for their work. For example, Marie received the Gegner Prize on three occasions, which carried a stipend of several thousand francs.

Pierre suffered multiple disappointments. In 1898, he applied for the chair in physical chemistry at Paris, but he lacked the typical French academic credentials, and much of his scientific work lay outside the boundaries of this discipline. Pierre seems to have lacked any capacity for self-promotion, finding the political aspects of academic life utterly repellent. To make ends meet, Pierre accepted a second position as tutor in the Polytechnic School.

Then in 1900 Pierre received an unexpected offer of the chair in physics from the University of Geneva. In contrast to the roughly 5,000 francs Pierre was receiving in Paris, he would earn 10,000 francs per year, and he would have a laboratory equipped to his specifications, as well as two full-time assistants. Moreover, Marie would receive a position in the same laboratory. Pierre accepted the offer, and in the summer, he and Marie journeyed to Geneva.

Later in the summer, however, Pierre decided to decline the offer. The move would require considerable effort to retool his teaching programme, and he and Marie would be forced to interrupt their research on radium and polonium for at least several months. Instead they would remain in Paris, where Pierre would accept a better paying position at the Sorbonne, and Marie would teach physics in a school for girls near Versailles. Although a chair had eluded them, at least their finances were eased.

In 1902, Pierre again applied for a professorship at the Sorbonne, this time in mineralogy. His important work with crystals, including the piezoelectric effect, would lend considerable weight to his candidacy. Jacques, Pierre's brother, held the chair in mineralogy at Montpellier. Yet again, Pierre had not attended the typical schools and therefore lacked support. Again, the chair went to another candidate.

Right: Marie and Pierre Curie in their laboratory, 1900.

NOMINATIONS

Perhaps to offer solace, Pierre was then
put forward as a candidate for election to
the French Academy of Science. Following
protocol, Pierre sought an audience with each
member. Later a journalist wrote of the great
discomfort the whole process caused him:

> To climb the stairs, have himself announced,
> say why he had come – all this filled the
> candidate with shame in spite of himself;
> but what was worse, he had to set forth his
> honours, state the good opinion he had of
> himself, boast of his science and his work –
> which seemed to him beyond human power.
> Consequently, he eulogized his opponent
> sincerely and at length, saying that he was
> much better qualified to enter the Institute.

Pierre was then proposed as a member of
the Legion of Honour. His nominator, the new
dean of his school, even reached out to Marie,
begging her to "use all your influence to
keep him from refusing it". Pierre's response
said it all: "Please be so kind as to thank the
Minister and to inform him that I do not feel
the slightest need of being decorated, but that
I am in the greatest need of a laboratory."
What mattered to Pierre was the work.

Right: The French Institute, where the Academy
of Science is located.

The Curies were working so hard that their health began to suffer. On several occasions, Pierre was seized by pains in his legs that landed him in bed. Marie was losing weight – according to her own notebook, her slight frame had thinned by 15 pounds. A colleague, alarmed at the change in Marie, wrote to Pierre, urging him to ensure that Marie started taking more rest and eating better, a prescription he thought advisable for Pierre, as well.

I have been struck, when I have seen Madame Curie at the Society of Physics, by the alteration in her appearance. I know very well that she is overworked. … She has not sufficient sources of resistance to live such a purely intellectual life as that which both of you lead. … It is necessary not to mix the scientific preoccupations continually into every instant of your life, as you are doing. You must allow your body to breathe. You must sit down in peace before your meals. You must not read or talk physics while you eat.

Marie's father was not doing well. Pierre's father lived with them, but Władysław had remained back in Poland, following his daughter's discoveries at a distance. Happily, Marie and Pierre had been able to visit on a few occasions in the preceding years. But in 1902, just days after receiving news that Marie had isolated radium, he suffered a gallbladder attack, underwent surgery to remove gallstones, and died. When Marie learned of his illness, she immediately boarded a train, but she arrived too late.

In May of 1903, Marie's dissertation, which had taken five years to complete, was at last accepted by the Sorbonne, and a month later, Marie defended it. As fate would have it, Ernest Rutherford happened to be visiting Paris at the same time, and he stopped by the Curies' lab, only to discover that they were at the ceremony. They were able to dine together that evening, and Rutherford later recalled the vial of radium that Pierre produced after dinner, glowing beautifully.

As Marie defended her dissertation, she was expecting another happy event: the birth of her second child. However, in August, five months pregnant, Marie suffered a miscarriage. Overcome with grief, she wrote to Bronya:

I had grown so accustomed to the idea of the child that I am absolutely desperate and cannot be consoled. … I had confidence in my organism, and at present I regret this bitterly, as I have paid dearly for it. The child – a little girl – was in good condition and was living. And I had wanted it so badly!

Soon thereafter bad news arrived from Poland: Bronya's young son had died of meningitis. Marie wrote that she could "no longer look at her little girl without trembling with terror".

Marie remained in poor health for much of the year, which would end with rumblings of a major prize from Stockholm.

Right: The front cover of Marie Curie's doctoral dissertation.

Série A, N° 445

N° d'ordre
1127.

THÈSES

PRÉSENTÉES

A LA FACULTÉ DES SCIENCES DE PARIS

POUR OBTENIR

LE GRADE DE DOCTEUR ÈS SCIENCES PHYSIQUES,

PAR

Mme SKLODOWSKA CURIE.

1re **THÈSE.** — RECHERCHES SUR LES SUBSTANCES RADIO-ACTIVES.

2e **THÈSE.** — PROPOSITIONS DONNÉES PAR LA FACULTÉ.

Soutenues le   juin 1903, devant la Commission d'Examen.

MM. LIPPMANN, *Président.*
BOUTY,
MOISSAN, } *Examinateurs.*

PARIS,

GAUTHIER-VILLARS, IMPRIMEUR-LIBRAIRE

DU BUREAU DES LONGITUDES, DE L'ÉCOLE POLYTECHNIQUE,
Quai des Grands-Augustins, 55.

1903

THE NOBEL PRIZE

THE NOBEL PRIZE WAS ESTABLISHED BY ALFRED NOBEL (1833–1896), A SWEDISH CHEMIST, ENGINEER, AND INDUSTRIALIST.

Left: Alfred Nobel.

Born in Stockholm, Nobel was the third of eight children of an impoverished inventor. After suffering several failures in business, the elder Nobel moved his family to St Petersburg, where he manufactured explosives. As the family's fortunes prospered, Alfred received a first-rate education and excelled in his studies.

In 1888, the death of Ludwig Nobel led several newspapers mistakenly to publish Alfred Nobel's obituary. One announced, "The Merchant of Death Is Dead." The opportunity to read his own obituary convinced Nobel that he should endeavour to leave a different legacy. Having neither married nor sired an heir, Nobel decided to leave the majority of his wealth to found the Nobel Prizes, which would be awarded for the first time in 1901.

According to Nobel's will, three prizes would be awarded for contributions in physical science (physics), chemistry, and physiology or medicine. The fourth would be awarded for literary work. And in a clear attempt to counteract his reputation as the man who did more than any other to develop means of killing large numbers of people, the fifth would be awarded for promoting international fraternity, reducing standing armies, and further peace (the Peace Prize).

AN EXPLOSIVE CAREER

As a young man, Nobel developed an interest in nitroglycerine, a powerful but highly unstable explosive, and he invented a detonator and the blasting cap. In 1864, his younger brother Emil was killed in an explosion while working with nitroglycerine. Three years later, Nobel invented dynamite, which retained nitroglycerine's immense explosive power but rendered it far safer. Nobel went on to invent other powerful explosives, such as gelignite.

Nobel became very rich. Two of his brothers, Ludwig and Robert, had remained in Russia, becoming oil barons who were sometimes referred to as the "Russian Rockefellers". Nobel's investments in their enterprises produced very handsome returns. Moreover, Nobel himself owned nearly 100 armaments factories around Europe, and he was able to amass more than 350 patents over the course of his lifetime, generating a considerable fortune.

Above: Alfred Nobel as a young man.

Left: Alfred Nobel's will.

Above: Nobel Prize medal.

Left: Marie's son-in-law Henry Labouisse Jr, who accepted the Nobel Peace Prize on behalf of UNICEF in 1965.

Of the nearly 1,000 people and organizations that have received a Nobel Prize, only a handful have been so honoured more than once. Linus Pauling (1901–1994) received Nobel Prizes in both chemistry and peace. John Bardeen (1908–1991) received the prize in physics twice, once for inventing the transistor and once for developing superconductivity. Frederick Sanger (1918–2013) received two chemistry prizes, one for elucidating the structure of insulin and one for a method of sequencing DNA.

Two organizations have received the Nobel Peace Prize multiple times. The International Committee of the Red Cross was so honoured three times, in 1917, 1944, and 1963. The United Nations High Commission for Refugees received it in 1954 and 1981. A Nobel Peace Prize awarded to the United Nations International Children's Educational Fund in 1965 was accepted by Marie's son-in-law, Henry Labouisse Jr (1904–1987), who was married to Marie's second daughter, Ève.

No family has received more Nobel Prizes than the Curies. Pierre and Marie received the Nobel Prize in Physics in 1903, and subsequently Marie received a Nobel Prize in Chemistry in 1911, becoming the only person to win Nobel Prizes in two different natural sciences. In 1935, Marie's daughter Irène and her husband Frédéric Joliot-Curie (1900–1958) received the Nobel Prize in Chemistry for proving that it is possible to transmute one chemical element into another.

NOBEL LAUREATES

IN 1903, WHEN FOUR MEMBERS OF THE FRENCH ACADEMY OF SCIENCES WROTE TO THE SWEDISH ACADEMY TO NOMINATE THEIR COLLEAGUES AND COUNTRYMEN FOR THE NOBEL PRIZE, THEY OFFERED ONLY TWO NAMES: HENRI BECQUEREL AND PIERRE CURIE.

Marie was left out, perhaps because the nominators thought that including her would compromise France's prospects of winning the award. There is no doubt that they were well aware of Marie's contributions. One, Gabriel Lippmann, had been one of her principal mentors.

Fortunately, an influential member of the Swedish Academy wrote to Pierre Curie to warn him that Marie might be excluded.

Pierre wrote back, "If it is true that you are seriously thinking of me, I very much wish to be considered together with Madame Curie in connection with our research on radioactive bodies." To this he added, "Don't you think it would be more satisfying, from an aesthetic point of view, if we were to be linked together in this manner?"

When the announcement was made, the 1903 Nobel Prize for Physics was shared by

Henri Becquerel and Pierre and Marie Curie. Marie became the first woman to win a Nobel Prize and remained the only one in the natural sciences until 1935, when her daughter Irène shared a Nobel with her husband. It is clear how new the Nobel Prize was at the time from a letter Marie wrote to her brother a few days after the award: "I do not know exactly what that represents; I believe it is about 70,000 francs."

The French press's reportage could hardly have been more hyperbolic. One weekly wrote:

> *Behold perpetual motion, the eternal sun, the supreme inexhaustible force has at last been found through the geniuses of the inventors Monsieur and Madame Curie, whose Nobel Prize fits them like a hand in glove.*

The French press rhapsodized over Marie, "a fair young woman, distinguished, slender in figure", and "a charming mother whose exquisite sensibility is accompanied by a spirit curious about the unfathomable". They even attributed an absurdly prideful quote to Pierre: "Sufferers from cancer, lupus, and paralysis will be healed by radium rays, no doubt. ... I am positive in what I assert. I never yet asserted anything that I could not prove, and I never will."

With such gross inaccuracies in wide circulation, it is no wonder that Pierre resented the burden of notoriety even more than Marie, writing to a colleague:

> *We have been pursued by journalists and photographers from all the countries of the world, and they have gone so far as to reproduce the conversation of my daughter with the maid and to describe the black and white cat who lives with us. Then we have received letters and visits from all the eccentrics, from all unknown inventors,*

Above: Marie and Pierre Curie's official 1903 Nobel Prize photographs.

Opposite: Royal Swedish Academy of Sciences.

and from all the unknowns in general – then we have had requests for money in great numbers, and finally collectors of autographs. … With all that, not an instant of tranquillity in the laboratory and a voluminous correspondence to take care of every evening. With this routine, I feel mindlessness invading me.

Decades later, the Curies' daughter Ève captured their predicament. On the one hand, they were delighted that their work was being recognized by the Swedish Academy. It meant a great deal to them to receive, among piles of congratulatory letters from people who did not understand their work well, a few enthusiastic messages from scientists they admired. And the

Above: 1903 Nobel Prize diploma of Pierre and Marie Curie.

70,000 francs in prize money would go a long way towards relieving their financial worries.

But this moment was also "perhaps the most pathetic of their lives". They were at their prime, an age when "genius, served by experience, could give its maximum". The Curies were being turned into inert idols at precisely the moment when they felt most poised to move forward. They had discovered radium and "astonished the world", but they also felt certain that new discoveries awaited them, and they desperately wanted to get on with their work.

At bottom, the press and public seemed to be entangling the Curies in a web of unreality. They had done great work, but now the world seemed intent on keeping them from their labours. They had been recognized by the world's scientific community, yet the University of Paris had not yet created a chair for Pierre. Marie knew that she needed to stay focused on what really mattered: her responsibilities as wife, mother, scientist, and teacher. Celebrity she refused.

Ève recounts an episode that captures this tension. While Marie was dining at the Élysée Palace with the French president, a woman approached her and asked if she would like her to present the President of Greece. Marie responded, "I don't see the utility of it." Surprised by the look of horror that spread across the woman's face, only then did Marie recognize the French president's wife. Blushing, Marie responded, "But naturally, I shall do whatever you please."

DISTASTE FOR NOTORIETY

The Curies were invited to visit Stockholm in December of that year, both to receive the award and to offer a public lecture. Not surprisingly, Pierre initially declined the invitation, citing the couple's teaching responsibilities and Marie's health – she had not entirely recovered from the miscarriage. Underlying these excuses was the Curies' distaste for ceremonies, public recognition, and time away from the laboratory that such an engagement would entail.

This aversion to notoriety is reflected in Marie's letter to her brother:

We are inundated with letters and with visits from photographers and journalists. One would like to dig into the ground somewhere to find a little peace. We have received an offer from America to go there and give a series of lectures on our work. They ask us how much we want. Whatever the terms may be, we intend to refuse. With much effort we have avoided the banquets people wanted to organize in our honour.

Ironically, the Curies' distaste for publicity may have only fanned the flames of popular fascination. In contrast to Becquerel, the Curies were unknown to the general public. Adding to the romance was the fact that they conducted their work in the most miserable of laboratories. And of course, there was the fact that Marie was a woman, a wife, and a mother, who happened to be married to her scientific collaborator.

RADIUM: REMARKABLE BUT DEADLY

RADIUM IS A REMARKABLE CHEMICAL ELEMENT. IT IS THE HEAVIEST ALKALINE EARTH METAL. ALL OF ITS 33 ISOTOPES ARE HIGHLY RADIOACTIVE. IN ITS MOST COMMON FORM, IT HAS A HALF-LIFE OF 1,600 YEARS, WHICH BOTH RAISES A PUZZLE AND EXPLAINS WHY THE CURIES FOUND IT IN PITCHBLENDE.

One might think that all the radium on earth should have disappeared long ago, but it is a natural product of the decay of uranium, which has a much longer half-life. Hence it is found with uranium in pitchblende, a by-product of silver mining.

Radium's fascinating scientific properties were matched and often exceeded by the public's fascination with the new element, which lasted for several decades. Minute quantities of radium began appearing in a variety of consumer products, including toothpastes, products to enhance male virility, and cosmetics such as facial creams. Thanks to radium, claimed advertisers, the term "radiant beauty" could take on a new and more literal meaning.

Shortly after the industrialization of radium's production, it began appearing in luminescent paints, which were used in watches, aircraft instrumentation, and clocks. In the days before small batteries, paint containing small quantities of radium made such surfaces glow

Below: Ad for Undark.

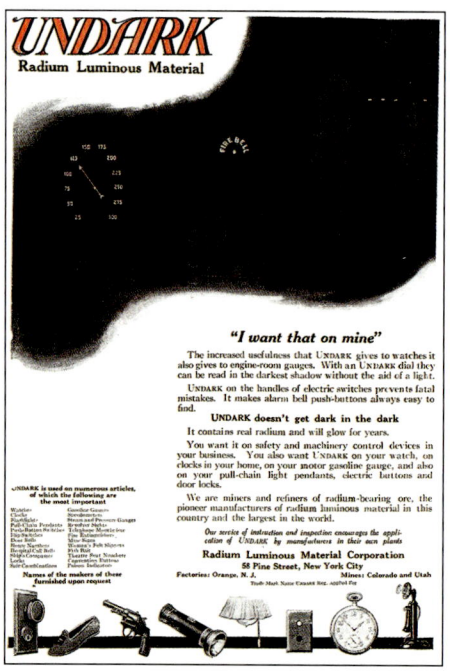

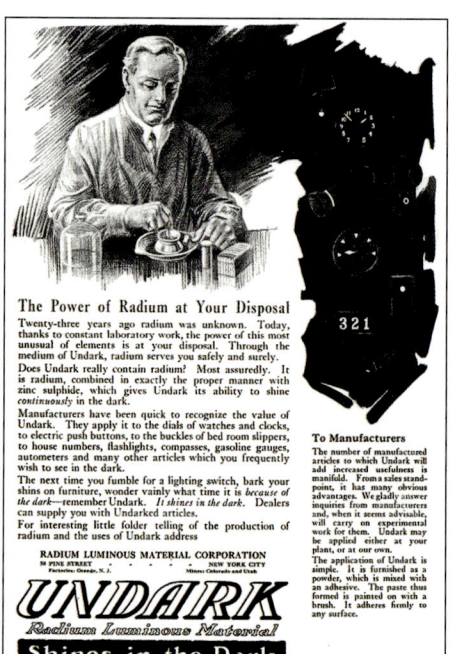

in the dark – a property especially valuable in military applications. An advertisement for one such paint, known as Undark, read as follows:

Manufacturers have been quick to recognize the value of Undark. They apply it to the dials of watches and clocks, to electric push buttons, to the buckles of bedroom slippers,

to house numbers, flashlights, compasses, gasoline gauges, autometers, and many other articles which you frequently wish to see in the dark. Next time you fumble for a lighting switch, bark your shins on furniture, wonder vainly what time it is because of the dark, remember Undark. It shines in the dark.

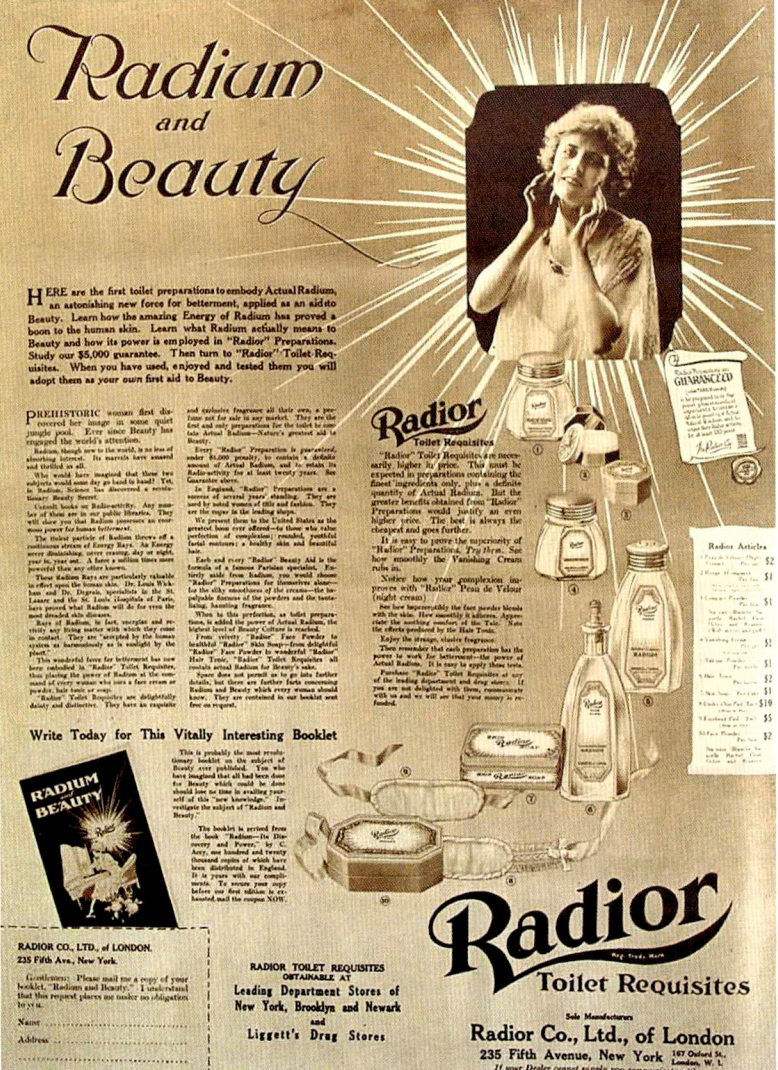

Left: Ad for radium beauty products.

One New Jersey company, the US Radium Corporation, employed 700 women, who received two cents for each watch dial they painted. Unfortunately, in an effort to achieve a fine point on their brushes, the workers often licked their tips of their paintbrushes. Even though the concentrations of radium were very low, assuming that a worker painted 250 dials per day and pointed her brush 15 times per dial, she might ingest thousands of times the maximum allowed dose.

Soon many of the young women – dubbed "radium girls" in the popular press – began developing health problems, including anaemia, decay of their jaws, and bone cancers. Eventually, the women brought lawsuits against their employers, alleging that their illnesses were traceable to radium exposure. Their claims received a boost when the

inventor of radium dial paint developed similar symptoms and died from cancer.

A Chicago newspaper described the scene as one of the radium girls appeared in a judicial proceeding to offer testimony:

A dark-haired, emaciated little woman painfully placed a brush to her lips yesterday and re-enacted on her deathbed the painful details of a process by which she and eleven other "doomed women" took fatal radium into their systems. Seven fellow sufferers sat around the bedside of Mrs. Catherine Donohue, 35, mother of two, and with faces drawn with emotional intensity saw her demonstrate the method of painting radium numerals on clock faces. … For Mrs. Donohue and perhaps others, the case is academic. Doctors have testified that she

ELIXIR OF LIFE

Members of the public also suffered from radium exposure. One product, Radiothor, was marketed as an "elixir of life". A wealthy Pittsburgh industrialist, Eben Byers, consumed a bottle or more of Radiothor every day, hoping to enhance his health and extend his life. Instead, his health declined rapidly, with lesions appearing on his skin and his bones undergoing decay. Ironically, he responded by increasing his consumption of Radiothor, only accelerating his decline.

Once the US amateur golf champion, Byers consumed approximately 1,400 bottles of Radiothor. When he died, one newspaper encapsulated his story in these terms: "The Radium Water Worked Fine Until His Jaw Came Off." A science journal reported that, at his death, his body contained the largest amount of radium ever found in a human being, and he was buried in a lead-lined coffin. The manufacture of Radithor ceased, but its producer soon turned to other radioactive products.

Right: Eben Byers in the 1920s.

probably will not live to see its outcome. … Thomas Donohue, husband of Mrs. Donohue, stood beside her bed and wept silently as he watched the scene.

In court, it emerged that company executives had taken steps to safeguard themselves from the dangers of radium but did little to protect their employees. Some hired agents to smear the reputations of the plaintiffs, even insinuating that their health problems were the result of syphilis. Eventually, however, the radium girls prevailed, soon spawning the first US occupational safety legislation. Sadly, the legal victory came far too late to save the lives of the radium girls.

No living organism requires radium, but it has been used in several legitimate biomedical contexts. For example, Nobelist Thomas Hunt Morgan (1866–1945) used radium to induce mutations in fruit flies, and Nobelist Herman Mueller briefly did the same before continuing his research with x-rays. Researchers at Johns Hopkins Hospital used radium in the treatment of cancer, sometimes inserting radium capsules directly into tumours. Today, safer isotopes are used.

The Curies were touched in many ways by "radium fever". An American dancer known for the spectacular light effects of her shows, having read that radium was luminous, soon showed up at their door, insisting that they provide her with radium to add to the effect. She even arranged to dance one evening at their house and soon became a friend of the family. Despite the friendship, however, the Curies never supplied her with the radium she sought.

Left: Radium dial painters working in a factory.

AFTER THE PRIZE

SHORTLY AFTER PIERRE AND MARIE RECEIVED THE NOBEL PRIZE, PIERRE WAS FINALLY OFFERED A NEW CHAIR IN PHYSICS AT THE UNIVERSITY OF PARIS. IT WOULD CARRY A SALARY OF 10,000 FRANCS PER YEAR, BUT ITS CREATORS HAD MADE NO PROVISION FOR A LABORATORY.

Pierre declined the offer. Confronted by outrage over the fact that such a celebrated scientist did not hold a chair, the university raised the ante, offering Pierre a laboratory, assistants and even creating a laboratory chief position for Marie. He accepted.

Pierre began arduous preparations to teach at the Sorbonne. He had to prepare new lectures, meet the needs of a new group of students, install new equipment, and continue to cope with a procession of admirers and well-wishers. Despite attaining this longed-for position, Pierre began to feel that, while constantly busy, he was accomplishing very little of significance. He had demonstrated the tremendous energy contained in radium, but it was Rutherford who would discern its atomic structure.

Pierre was suffering not only from a host of new responsibilities but from increasing fatigue. He accepted the Nobel Prize lecture, under the condition that it be postponed. He wrote to a friend:

> I have had to give up going to Sweden. We are, as you see most irregular in our relations with the Swedish Academy; but, to tell the truth, I can only keep up by avoiding all physical fatigue. And my wife is in the same condition; we can no longer dream of the great workdays of times gone by.

Later in 1905, Pierre was finally feeling well enough to journey to Stockholm, accompanied by Marie, to deliver the Nobel Prize lecture. Entitled "Radioactive Substances, Especially Radium," it was delivered by Pierre, while Marie sat in the audience. This approach reflected Marie's earnest evaluation of their respective merits. Describing her admiration for her husband, she wrote:

> He was all I could have dreamed at the moment of our union, and more. My admiration for his exceptional qualities, on a level so rare and high, constantly increased, so that he sometimes seemed to me like an almost unique being, by his detachment from all vanity and from those pettinesses which one finds in one's self and in others, and which one judges with indulgence, though not without aspiring to a more perfect ideal.

During his speech, the modest Pierre mentioned his wife's contributions to their work twice as often as his own. He considered the implications of the discovery of radium for a broad array of fields, including chemistry and biology. And, in his conclusion, he speculated on the role the phenomenon of radioactivity would come to play in the course of human affairs:

BIRTH OF SECOND DAUGHTER

The Curies were buoyed by the prospect of the birth of their second child, Ève, who arrived in December of 1904. As the birth approached, Marie became increasingly gloomy, perhaps from memories of the miscarriage she had suffered. After the birth, however, she recovered quickly. She did not attempt to nurse the baby herself, instead immediately hiring a wet nurse. With time, she began to suspect that Irène was jealous of the attention showed to her younger sister.

Right: Irène Curie holding her younger sister, Ève.

*One can imagine that in criminal hands
radium could become very dangerous and
one can ask if humanity is at an advantage
in knowing nature's secrets, if it is mature
enough to make use of them or if this
knowledge might not be harmful to it. The
example of the discoveries of Nobel is a case
in point; powerful explosives have allowed
men to do admirable work. They are also a
terrible means of destruction in the hands of
great criminals who lead people into war. I
am among those who think with Nobel that
humanity will derive more good than bad
from new discoveries.*

Perhaps Pierre's speculations on the future
of radioactivity were grounded in part on
the difficulties it was causing him. It had
become apparent that his health was in a state
of decline, which his physicians blamed on
rheumatism. His hands had become so stiff
that he had difficulty even dressing himself, let
along performing laboratory experiments.
He lacked the stamina of days gone by, and
from time to time the pains in his legs grew so
severe that he could find no rest, even in bed.

Marie began to worry at the possibility that
her husband might never be well again, and
Pierre wrote to a friend:

*I am neither very well nor very ill. But
I get tired easily, and I no longer have more
than a very feeble capacity for work. My
wife, on the contrary, leads a most active
life, between her children, the school, and
the laboratory. She does not lose a minute
and attends much more regularly than I do
to the progress of the laboratory, in which
she passes the greater part of her day.*

What seems to have troubled Pierre most of
all, however, was a lack of productivity. He
lamented his distress in a letter to a friend:

*We continue to lead the same life of people
who are extremely occupied, without being
able to accomplish anything interesting. It
is now more than a year since I have been
able to engage in any research, and I have
no moment to myself. Clearly, I have not
yet discovered a means to defend ourselves
against the frittering away of our time,
which is nevertheless extremely necessary.
Intellectually, it is a question of life or death.*

Opposite and above: Pierre Curie in 1906.

THE DEATH OF PIERRE

THE DATE WAS 19 APRIL 1906, A RATHER DARK AND RAINY DAY IN PARIS. PIERRE PLANNED TO ATTEND A LUNCHEON OF THE ASSOCIATION OF PROFESSORS OF THE SCIENCE FACULTIES AND THEN CORRECT THE PROOFS OF AN ARTICLE AT THE PUBLISHERS.

Usually sceptical of such organizations, Pierre regarded the APSE as an opportunity to position the sciences at the core of education and to ensure that scientists were hired based on merit, not political considerations.

Pierre was so highly regarded that he had been elected president of the organization, and he relished the opportunity it afforded him to inject more reason into science. The lively discussion at the meeting prompted Pierre's former student, Paul Langevin, to remark sometime later that he had never seen Pierre more animated and upbeat than at that meeting. This spirit was reinforced by the presence of Joseph Wierusz-Kowalski, who had introduced the Curies to one another 12 years prior.

After the meeting and a failed attempt to visit his publisher, whose offices were closed due to a strike, Pierre set forth into the downpour, protected from the rain by his umbrella. As he crossed the Rue Dauphine at the Quai de Conti, one of the busiest intersections in Paris, he unexpectedly encountered a large, horse-drawn wagon. He reached out to one of the horses to steady himself on the slick pavement, but the horse reared, and Pierre fell.

As his daughter Ève would later describe the scene,

Pedestrians cried, "Stop! Stop!" The driver pulled on the reins, but in vain: the team of horses kept up. Pierre was alive and unhurt. He did not cry out and hardly moved. His body passed between the feet of the horses without even being touched, and then between the two front wheels of the wagon. A miracle was possible. But the enormous mass, dragged on by its weight of six tons, continued for several metres more. The left back wheel encountered a feeble obstacle which it crushed in passing: a forehead, a human head. The cranium was shattered, and a red, viscous matter trickled in all directions in the mud: the brain of Pierre Curie.

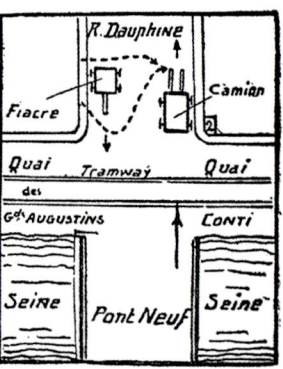

Left: Sketch of location of Pierre's death.

Pierre was killed instantly. Immediately, a crowd began to gather. It was obvious to all that no aid could be rendered to the fallen man, and people started surrounding the wagon driver, threatening him for his actions. Police detained him and, after extensive questioning, determined that he had not been at fault. Later, Pierre's friends would put much of the blame on his own absent-mindedness, his distracted way of going about the ordinary activities of life.

When word reached the university, the task of informing Pierre's family fell to the dean of sciences, who trudged to the Curies' residence to deliver the news. When Pierre's father came to the door and saw the grieved expression on his visitor's face, he blurted out, "My son is dead." Marie, as it turned out, had gone out for the day with Irène. They waited three hours for her return, during which time Pierre's father muttered repeatedly, "What was he dreaming of this time?"

Marie arrived home in good spirits, but immediately recognized that something was amiss. As she later described the scene in a journal entry addressed to Pierre,

I enter the room. Someone says, "He is dead." Can one comprehend such words? Pierre is dead, he who I had seen leaving looking fine this morning, he who I expected to press in my arms this evening, I will only see him dead and it's over forever. I repeat your name again and

Below: Depiction of the death of Pierre Curie.

always, "Pierre, Pierre, Pierre, my Pierre,"
alas that does not make him come back,
he is gone forever, leaving me nothing but
desolation and despair.

The next day, newspapers around the world carried stories announcing the great scientist's death. Yet they were largely overshadowed by accounts of the great earthquake that had struck the city of San Francisco the very same day, killing many hundreds. Marie arranged for Irène to stay with a friend and telegraphed her relatives in Warsaw, "Pierre dead result accident". The next day, Pierre's brother Jacques arrived, and for the first time Marie was observed to break down in tears.

Initially Marie told Irène that her father had sustained an injury to his head and was resting, but the day after the burial she told her the truth. At first, the girl did not understand, and let her mother leave without explaining anything. But Marie later learned that she had wept and asked to go to her home. There, according to her mother, she cried a great deal,

but soon she went back to her friends, reluctant to speak of her father.

Little Ève was too young to understand, and it was only later that she began to grasp that her father was gone forever.

Describing the effect of Pierre's death on Marie, Ève later wrote,

It is commonplace to say that a sudden catastrophe may transform a human being forever. … [But] Marie Curie did not change from a happy young wife to an inconsolable widow. The metamorphosis was less simple and more serious. The inner tumult that lacerated Marie, the nameless horror of her wandering ideas, were too virulent to be expressed in complaints or confidences. From the moment when those three words, "Pierre is dead," reached her consciousness, a cape of solitude and secrecy fell on her shoulders forever. Madame Curie, on that day in April, became not only a widow, but at the same time a pitiful and incurably lonely woman.

THE FUNERAL

Following Pierre's wishes, Marie decreed that the funeral should lack all spectacle. There would be no procession and no speeches, and Pierre would be buried simply in the family plot where his mother already rested. During the brief ceremony, Marie remained silent, clutching her father-in-law's arm except when a sheaf of flowers was brought near. Then she reached out, took them, and one by one scattered them on the coffin.

Of the scene at the burial, Marie later wrote,

I put my head against the coffin and in great distress, I spoke to you. I told you that I loved you and that I had always loved you with all my heart. … I promised that I would never give another the place that you occupied in my life and that I would try to live as you would have wanted me to live.

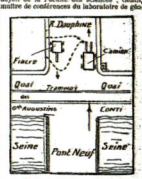

Left: Newspaper front-page story announcing the death of Pierre Curie, *Le Matin*, 20 April 1906.

DARK DAYS

MARIE FOUND LIFE WITHOUT PIERRE UNBEARABLE, AND THOUGH SHE HAD RESOLVED NOT TO TAKE HER OWN LIFE, SHE SOMETIMES FOUND HERSELF WISHING THAT ONE OF THE MANY CARRIAGES ON THE STREETS OF PARIS WOULD GRANT HER THE FATE OF HER BELOVED.

Things that would have brought light into her life only reminded her of the loss and deepened the gloom – the blossoming of spring, an amusing word from one of the children, or a moment of immersion in scientific research.

If there was any hope for her, it seemed to lie in the laboratory, where she and Pierre had spent most of their married time together. This feeling increased as the university grappled with the question of what to do with Pierre's

Above: A discouraged-looking Marie Curie alone in her laboratory.

chair. Today it might seem obvious that the chair should be given to his scientific partner, but at the time the idea that a woman would teach at the Sorbonne, let alone hold a professorship, was without precedent.

Unable to commit fully to the idea, the university decided to leave the chair vacant and offer Marie the position of director of the laboratory. Her friends urged her to accept it, partly because she was by far the best qualified person to carry on the research they shared, partly because it would keep the name Curie attached to the work, and partly out of concern for Marie's own welfare. They could see that the best hope for Marie's recovery lay in work.

A year later, Marie wrote to one of her friends,

> *My life is upset in such a way that it will never be put right again. I think it will always be like this, and I shall not try to live otherwise. I want to bring up my children as well as possible, but even they cannot awaken life in me. They are both good, sweet, and rather pretty. I am making great efforts to give them a solid and healthy development. When I think of the younger one's age, I see it will take twenty years to make grown persons of them. I doubt if I last so long, as my life is very fatiguing, and grief does not have a salutary effect upon strength and health.*

Marie received great support from Pierre's father. A stoic resolve made him carry on with life, talking and laughing as he always had. Seeing his courage, Marie began to feel ashamed of her mourning. The children loved their grandfather, who spent more time with them than their mother, serving as both their

Below: Single parent Marie Curie with her daughters Irène and Ève.

TEACHING AT THE SORBONNE

In the autumn of 1906, Marie became the first woman ever to teach a course at the Sorbonne, one of the oldest universities in the Western world that had existed since the thirteenth century. Her first lecture turned out to be a sensation, with hundreds of people queuing up outside the door, hoping to get a seat. Every seat in the amphitheatre was occupied, and when Marie walked into the room at the appointed hour, the audience erupted in applause.

Those present may have hoped that Marie would offer some words of tribute or express sorrow at the passing of her husband, but instead she launched into a review of recent progress in the field of physics. At 39 years of age, Marie felt that, except for her closest family members, the social dimension of her life had closed forever. If she spoke, it was only about work or the education of her daughters. Otherwise, she had nothing to say.

Right: Marie Curie, first woman professor at the Sorbonne.

teacher and their playmate. Marie was fully cognizant of the difference he had made, and when in 1909 his heart began to fail, she spent spare moments at his bedside, tending to him.

After the old man died in 1910, Marie was left to raise the girls alone. Her top priorities as a parent were to instil independence and hardiness in her children. They were sent out on long walks, regardless of the weather. She encouraged such pursuits as gymnastics, having bars and rings installed in the garden. She had no patience for fears of any kind – of thunderstorms, diseases, or bogeymen. Pierre's accidental death would not be allowed to render his daughters timorous.

In fact, Marie rarely referred to Pierre, and it was only with difficulty that she spoke his name. Nor did she talk of the circumstances of her own childhood – the deaths of her sister and mother, the financial straits, and the persecution her family suffered at the hands of the Russians. Marie had long ago abandoned her mother's faith, but she made no effort to prevent her daughters from adopting whatever religious tenets they wished.

She had no use for conventional schools, which seemed to her merely to confine pupils in poorly lit and ventilated rooms when they should be roaming about free. So she sought the advice and support of many of her academic colleagues and founded a teaching cooperative, in which children would be instructed in subjects such as physics, maths, and literature by university faculty members. Many of the students quickly developed a deep love of learning and went on to become scientists themselves.

Ève later credited the two years of cooperative education with instilling in the girls a taste for hard work, an indifference toward money, and an independent spirit that would enable them to carry on in difficult circumstances without need of help. And yet decades later, upon Marie's death, all the letters her daughters had ever written her were found among her few treasures. Underneath her firm and independent exterior, Marie harboured a desperate longing for affection.

Left: Marie Curie posing with some of her female students

THE FRENCH ACADEMY

IN THE YEARS LEADING UP TO 1910, MARIE HAD WORKED HARD AND RECEIVED MANY ACCOLADES. SHE WAS CLOSE TO PURIFYING RADIUM AND HAD PRODUCED TWO MAJOR WORKS ON RADIOACTIVITY.

The first, published in 1908, was a 600-page composition called *Works of Pierre Curie*. Two years later, she published the 1,000-page *Treatise on Radioactivity*, which summarized nearly everything that was known about the phenomenon that Marie herself had named little more than a decade before.

For her labours, she had received more recognition than any woman in the history of the natural sciences. She had been the first (and at that time still the only) woman to receive a Nobel Prize. She had been appointed to the

International Radium Standards Committee. And she had been elected to the American Philosophical Society and the national academies of Poland, Czechoslovakia, Russia, the Netherlands, and Sweden. She was also world famous.

So it seemed to many quite appropriate that Marie should be elected to the French Academy, the same organization that years before had – after early disappointment – eventually welcomed Pierre. Officially established by Louis the Fourteenth in 1666,

Right: Mémoires de l'Academie des Sciences, volume 46, 1903, which contains the first printing of Henry Becquerel's landmark treatise on radioactivity.

Far right: Marie Curie's book on her husband, including autobiographical notes.

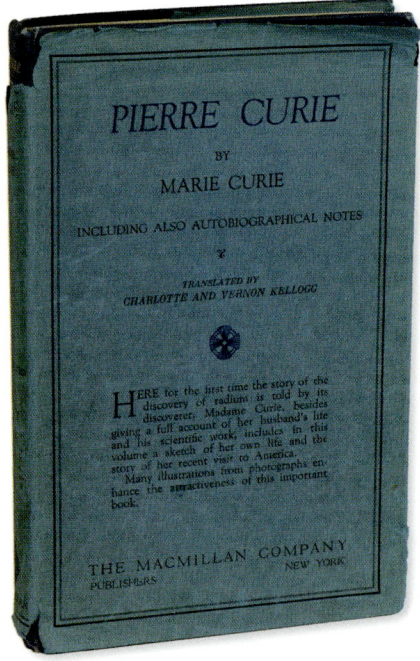

the Academy has been at the forefront of many scientific achievements in the 1600s and 1700s. The Academy of Sciences was one of five academies in the French Institute, and election to the Academy was the highest honour France could accord a scientist.

In 1910, a seat among the Academy's "Immortals" became vacant with the death of an eminent chemist. It was relatively easy for Marie's friends to make the argument that, based on her scientific merits, she deserved to be elected. They could also argue that membership would enable her to provide the Academy with unequalled expertise in the burgeoning field of radioactivity. Marie was soon convinced that she should allow herself to be nominated.

Although many candidates were formally under consideration, Marie's principal rival was Édouard Branly (1844–1940), a much older man who would be nominated three times for the Nobel Prize but never received it.

The contest between Marie and Branly to be elected to the Academy was framed along many lines that had little or nothing to do with the merits of their scientific work. One was religion. Branly was seen as a loyal Catholic, having chosen a position at the Catholic Institute over the University of Paris, while Marie was portrayed by her opponents as either largely indifferent to religion, antagonistic to it, or, in at least one case, as a closeted Jew.

Another point of contention was time. Branly, at age 67, was seen by many as nearing the end of his life, and he had previously been denied admission to the Academy. Marie, by contrast, was only 43 years old, and some argued that she would have many additional opportunities to be elected in years to come. Ironically, Branly ended up living to the age of 95, dying in 1940 and thus outliving Marie by six years.

Also at issue was their scientific merit. At first glance, Marie's contributions clearly

exceeded Branly's, and there should have been little debate. But some argued that Pierre had been primarily responsible for the work that Marie and he performed, and that her role had been that of an assistant. Against such claims Marie's proponents cited Pierre's own testimony, as well as the fact that Marie had remained quite productive even after his death.

But the single greatest point of contention between the two candidates was sex. Some pointed to the fact that in over 200 years of existence, the French Academy had never elected a woman. They appealed to tradition and argued that the Academy should remain true to its own history and respect the judgement of its past members, who had never seen fit to elect women. To admit a woman, some argued, would introduce a change in the Academy from which it would be difficult to recover.

Others, including more than a few women, argued that it was simply inappropriate for a woman to take her place in a scientific academy. Science, engineering, mathematics – these were the proper purview of men, and women should remain focused on characteristically feminine activities such as the cultivation of grace and beauty and the enrichment of domestic life. If women joined the Academy, they argued, it would trivialize the quality of discourse and distract the male members.

On the day of the vote, attendance spiked. Members who would not have otherwise attended made it a point to be present, and crowds gathered outside. Initially only members were allowed in the hall, but the president bowed to pressure to admit those outside, though only men. The only woman in attendance was a newspaper editor, and she gained admittance only after her colleagues in the press pleaded on her behalf.

To win election, a candidate would need to receive 30 votes. On the first ballot, Branly received 29 votes and Marie 28, with a single

ÉDOUARD BRANLY

A physicist and professor at the Catholic Institute, Branly was best known as the inventor of the "coherer", a device for use in radio transmission. After its introduction in the 1890s, for about a decade the coherer became the preferred means for receiving radio signals.

Many in France had assumed that when the Nobel Prize for Physics was awarded in 1909 for contributions to the development of wireless telegraphy, Branly would be among the recipients. The importance of his contributions was widely recognized: when Guglielmo Marconi (1874–1937) conducted his first radio transmission across the English Channel in 1897, he acknowledged Branly by transmitting, "Mr. Marconi sends to Mr. Branly his regards over the Channel through the wireless telegraph, this nice achievement being partly the result of Mr. Branly's remarkable work."

Left: Édouard Branly, Marie's principal 1911 rival for a seat in the French Academy.

Right: Guglielmo Marconi, recipient of the 1909 Nobel Prize in Physics for the development of radio.

Left: The Académie des Sciences in Paris was formed in 1666.

vote to a third candidate. On the second ballot, there were 30 votes for Branly and only 28 for Marie. Edouard Branly had been elected to the French Academy. Despite the loss, some regarded the outcome as a victory for Marie, presaging a day when candidates would no longer be judged by their sex.

Others were incensed at the result, seeing it as a disgrace not only for the French Academy but the whole nation of France, especially when so many other nations' science academies had already elected women members – Marie among them. Yet if Marie was crushed, she displayed no outward sign, and her friends and colleagues seem to have been more upset about it than she. Moreover, she may have been comforted by the precedent of Pierre, who had also been initially rejected.

Un Tournoi Académique : Une femme entrera-t-elle à l'Institut ?

Mᵐᵉ CURIE OU M. BRANLY ?

C'est aujourd'hui que l'Académie des Sciences doit nommer un successeur au fauteuil de M. Gernez. Les deux candidats sont, on le sait, Mᵐᵉ Curie et M. Branly. Les titres de l'une et de l'autre sont considérables. Mᵐᵉ Curie a collaboré avec son mari à la découverte du radium, M. Branly est l'inventeur de la télégraphie sans fil. Lundi dernier, la section de physique a désigné Mᵐᵉ Curie en première ligne par 3 voix contre 2. L'Académie des Sciences va-t-elle ratifier ce choix ?

Opposite: Marie Curie in her laboratory in the Sorbonne, c. 1908.

Below: "An Academic Tournament: Will a Woman Enter the Institute?", article in *Excelsior*, 9 January 1911.

PAUL LANGEVIN

THE STORY OF PAUL LANGEVIN INTERTWINED WITH THAT OF THE CURIES IN MANY WAYS. WHEN HE WROTE HIS PHD THESIS ON IONIZED GASES, HIS SUPERVISOR WAS PIERRE CURIE.

When in 1925 the Curies' daughter Irène completed her doctoral dissertation on the decay of polonium, one of the elements discovered by her parents, her supervisor was Langevin. As we shall see, the Langevin–Curie association took on an even more intimate character about four years after Pierre's death.

Langevin was born in Paris in 1872 and educated in some of France's best schools. At the end of the nineteenth century, he travelled to Cambridge, where he studied with J. J. Thomson, who was then well on his way to discovering the electron. After he returned and completed his PhD in 1902, he was invited to lecture on physics at the College of France. When Pierre left the School of Industrial Physics and Chemistry for the University of Paris, Langevin assumed his teaching duties.

Langevin made a number of important contributions to physics. Building on Pierre's work exploring the relationships between the magnetic properties of materials and their temperature, Langevin suggested that there was a competition between magnetism and random thermal motion. Through his work, he became one of the first scientists to explain the macroscopic properties of a substance in terms of the properties of atomic particles.

Langevin also theorized that the mass and energy of atomic particles were associated, although when Einstein published his theory of special relativity, Langevin acknowledged the precedence of Einstein's theory and became one of its most ardent boosters. Many years later,

Left: Paul Langevin as young man.

Right: Paul Langevin, early 1940s.

on the occasion of Langevin's death, Einstein would write, "It seems to me certain that [Langevin] would have developed the special theory of relativity if it had not been done elsewhere; he had already recognized clearly its main points."

Langevin also made important contributions to the understanding of Brownian motion, the random motion of particles in a fluid. Einstein had described such motion at a macroscopic level using thermodynamics, but it was Langevin who looked at the motion of

an individual particle, describing its collisions in terms of random forces. Today Langevin's equation is used to describe phenomena in many other fields, including chemistry, biology, and even economics.

Langevin was also deeply engaged politically. As a young man, he was active in the French League of Human Rights, and he was a strong supporter of the League of Nations. In 1923, he marched with Einstein in a Berlin demonstration for human rights. A committed pacifist, he spoke out strongly

SONAR

Perhaps Langevin's best-known contribution was the development of sonar. The most famous disaster of the age had been the 1912 sinking of the ocean liner *Titanic*, as the result of its collision with an iceberg, resulting in the deaths of over 1,500 people. With the widespread introduction of submarines in World War I, seafaring had become quite dangerous. How could ships defend against an enemy they could not see? Some means needed to be devised to detect large underwater objects.

Light did not work well – even the most powerful beams were quickly attenuated by seawater. Sound seemed a better candidate for use in detection, but the British invention of the hydrophone, an underwater microphone, provided little information on the location or distance of underwater objects. Working with an engineer, Langevin realized that Jacques and Pierre Curie's long-ago discovery of the piezoelectric effect provided the key to solving the problem.

Specifically, Langevin and a colleague produced a mosaic of quartz crystals, glued together between two steel foils. When an electric current passed through the device, it emitted sound waves. When the sound waves echoed back from an underwater object, they caused the crystal to vibrate, generating a current. Refinements of this technique could then be used to determine the location, distance, and even speed and direction of underwater objects.

SONAR (SOund NAvigation and Ranging) underwent further development after the war, and the underlying principles were later applied in a variety of fields. For example, RADAR, which relies on the reflection of radio waves, can be used to track aircraft and weather systems. Langevin also laid the groundwork for medical ultrasound, which relies on the reflection of high-frequency sound waves to image internal structures such as the liver and gallbladder, the thyroid gland, and the developing foetus.

Above: Robert Boyle. Boyle and Paul Langevin were two pioneers of submarine detection.

Right: Cross-sectional view of a form of quartz transducer designed by Boyle in 1917.

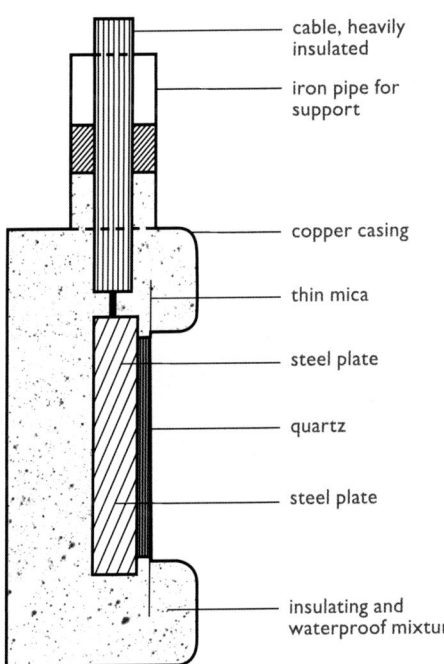

cable, heavily insulated

iron pipe for support

copper casing

thin mica

steel plate

quartz

steel plate

insulating and waterproof mixture

against fascism, but eventually came to believe that stiffer measures were needed to contain Hitler. At the end of his life, he joined the Communist Party.

Having participated with Marie and other colleagues in her short-lived cooperative for the education of their children, Langevin was also a dedicated educator. He believed that science needed to be widely shared with the public, and he advocated for a growing presence of science in school curricula. After World War II, he chaired an influential commission on the reform of the French educational system, although its recommendations were not implemented until after his death in 1946.

In addition to Irène Curie, Langevin's students included the quantum physicist and Nobel laureate Louis de Broglie, who proposed that all matter has wave properties, and Léon Brillouin, who made important contributions to quantum mechanics, solid state physics, and information theory.

Below: Langevin (centre) surrounded by other physicists: (left to right) Albert Einstein, Paul Ehrenfest, Kamerlingh Onnes and Oierre Weiss.

Left: Paul Langevin
and his wife Emma.

Langevin was elected to the Royal Society, and he also received both the Hughes Medal and the Copley Medal for his outstanding contributions to science.

In 1898, Langevin married the former Emma Jeanne Desfosses (1874–1970), the daughter of a ceramicist, with whom he would have four children. The fraught nature of their marriage, which seems to have been troubled almost from the start, had Langevin contemplating divorce within just a few years. The strife within the household would have profound repercussions not just for Langevin but also for the Curie family, and especially Marie.

THE LANGEVIN AFFAIR

TO SAY THAT LANGEVIN'S MARRIAGE WAS UNHAPPY WOULD BE AN UNDERSTATEMENT. FRIENDS REPORTED THAT LANGEVIN OFTEN APPEARED MISERABLE, THAT HE SOMETIMES SHOWED UP AT WORK WITH BRUISES AND CUTS ON HIS HEAD, AND THAT DURING SOCIAL DINNERS HIS WIFE WOULD SO SEVERELY UPBRAID HIM THAT LANGEVIN WOULD LEAVE THE TABLE.

Whether Langevin ever resorted to violence is unclear, but we do know that Madame Langevin complained about her husband's "harshness".

Within a year or two after Pierre's death, Marie became part of a close circle of friends in whom Langevin felt he could confide. By 1910, these confidences had blossomed into love, and the two rented an apartment near the university where they could be alone together. When apart, they also maintained a correspondence, the hazards of which Langevin must have perceived – many years earlier, his wife had intercepted a letter from his mother, expressing concerns about the state of his marriage.

As it turned out, Madame Langevin had obtained one of their letters, and she openly expressed her determination to make it public. This would be made easier by the fact that her brother was a newspaper editor. In addition, Madame Langevin threatened to kill Marie. Marie, overcome by worry, fell ill, and travelled with her daughters to the coast, where she hoped to recuperate. Though absent from Paris, she apparently envisioned a future with Paul Langevin.

She wrote to him:

There are quite deep affinities between us that require only favourable circumstances to develop. We have some hint of it in the past, but it didn't become fully apparent until we found ourselves face to face, I in mourning for the beautiful life that I had made which collapsed in such a disaster, and you with your sense that, in spite of sincere efforts, you had completely missed out on the family life in which you sought such great joy. The instinct which led us together was very powerful, since it helped us to overcome so many sad impressions about the different ways we had understood and arranged our private life. What couldn't come out of this desire, instinctive and natural and so compatible with our intellectual longings, to which it seems so well suited? I believe that we could derive everything from it: good work together, good solid friendship, courage for life, and even beautiful children of love.

It seems clear that Marie hoped Langevin would find the courage to divorce his wife. She portrays Madame Langevin as an inexorably domineering force who would keep her husband perpetually unsettled, consigning him to perpetual misery and interfering with his work. From Marie's point of view, Madame Langevin appeared a woman incapable of constraining her jealousy and rage and a schemer prepared to go to any lengths to keep her household.

In fact, Marie herself seems to have been consumed by jealousy. She urged Langevin to stay away from the marital bed. She suspected that Madame Langevin may try to become pregnant again, foreseeing that the combination of an impending birth and the tearful scenes that would attend it would undermine any resolve he might have to leave. Marie complained that the very thought of Langevin with his wife made it impossible for her to sleep.

Marie and Langevin made the mistake of keeping their letters to each other in their apartment in the city. Madame Langevin apparently hired someone to gain entry to the apartment and take the letters. Soon after this had been accomplished, her brother visited Marie and told her that Madame Langevin had the letters in her possession and intended to make them public. To Marie, it was a clear case of blackmail – end all association with Langevin or endure unthinkable public disgrace.

Right: Marie Curie and Paul Langevin, Paris, early 1910s.

Marie and Langevin seem to have seen less of each other in the ensuing months, although they were together at the Solvay physics conference in Brussels in 1911. A famous photograph from the conference shows Marie Curie seated at a table flanked by physicists on either side, while behind her stands Albert Einstein, and next to him, Paul Langevin. It is possible that news of the conference provoked Madame Langevin to make good on her threat.

The day after the conference ended, a Paris newspaper carried a front-page story about the affair. The reporter described a conversation at the Langevin home with the mother of Madame Langevin:

The widow of Pierre Curie, the great scientist, who collaborated in the discovery of radium, who is a professor at the University of Paris, who almost gained entry to the Academy of Sciences, the celebrated, the famous Madame Curie has carried off the husband of my daughter, the father of my little grandchildren.

Marie attempted to rebut these claims, but only partially. She protested that the conference in Brussels had been so engrossing that it consumed all her energies, leaving no time for anything else. She did not, however, claim that rumours about an affair with Langevin were entirely unfounded. In fact, she couldn't do so, both because she knew that saying so would be false, and because Langevin's mother-in-law had already revealed that she and her daughter had the letters in their possession.

Madame Langevin also appeared in the press. One account featured an interview, in which she was described as a "woman in tears", who wished to avoid a scandal and would never have made matters public if it

were only a question of an affair. She had hoped that she would have been able to draw her husband back to her. If she were the jealous madwoman some alleged, she would have been shouting from the rooftops, but she kept silent, out of her duty as a wife and mother.

Regardless of what those closest to the situation knew to be the truth, the story as it developed in the newspapers fostered sympathy for Madame Langevin and resentment toward Marie. Madame Langevin was merely trying to preserve her marriage, help her husband to see reason, and protect her children. Marie, on the other hand, a coldly rational foreigner, was engaged in a self-serving scheme to wreck a home and destroy a great scientist.

Above: December 1911 *Times Herald* headline announcing that a divorce had been granted to Madame Langevin.

THE SOLVAY CONFERENCE

BORN IN BELGIUM, CHEMIST AND INDUSTRIALIST ERNEST SOLVAY (1838–1922) PATENTED A PROCESS FOR PRODUCING SODIUM CARBONATE FROM SEA WATER AND LIMESTONE. KNOWN AS THE SOLVAY PROCESS, ITS PRODUCT IS USED IN THE MANUFACTURE OF GLASS, PAPER, SOAPS, AND DETERGENTS.

Left: Belgian chemist and industrialist Ernest Solvay, founder of the Solvay Conference.

Right: Hendrik Lorentz, Dutch 1902 Nobel Laureate in Physics and first chair of the Solvay Conference.

His first factory was in Belgium, but there are now plants using the Solvay process all around the world. As a result, Solvay made a large fortune, which he used for a variety of philanthropic purposes.

One of Solvay's most enduring contributions was the creation of the Solvay conferences, the first of which was held in Belgium in 1911. As Solvay envisioned the conference, it would serve not merely as a forum for the presentation of scientific papers but primarily focus on unresolved scientific problems. Its first chair was Hendrik Lorentz (1853–1928), a Dutch physicist who shared the second Nobel Prize in Physics in 1902. Lorentz's command of multiple languages proved instrumental in moderating the sessions.

THE EARLY CONFERENCES

The first of the invitation-only conferences focused on the growing tension between classical physics and quantum theory. Among those present were Max Planck (1858–1947), whose discovery that energy is transferred in discrete amounts or quanta would win him the 1918 Nobel Prize in Physics; Ernest Rutherford (1871–1937), who proposed the concept of half-lives of radioactive materials, winning the Nobel Prize in Chemistry in 1908; and the second-youngest attendee, Albert Einstein. Also present were Marie and Langevin.

The best known of all the Solvay conferences was the fifth. The last one chaired by Lorentz, it was held in 1927, and the theme was electrons and photons. A photo of the attendees is often described as "the most intelligent photograph ever taken", largely because no fewer than 17 of those present had already won or would win Nobel Prizes. Only one woman is pictured, and that of course is the only attendee to win two Nobel Prizes, Marie Curie.

Below: Photograph of participants in the 1927 Solvay Conference, often referred to as "the most intelligent photograph ever taken".

Among those present at the fifth conference, in addition to Lorentz, Planck, Rutherford, Einstein, Marie, and Langevin, were:

Erwin Schrödinger (1887–1961) – Austrian physicist who helped to advance quantum theory, co-won the 1933 Nobel Prize in Physics, and wrote the book *What Is Life?* which influenced the work of biologists such as James Watson, co-discoverer of the structure of DNA.

Wolfgang Pauli (1900–1958) – Swiss-American physicist who developed the Pauli exclusion principle, which says that two subatomic particles in the same atom cannot occupy the same quantum state simultaneously, for which he received the 1945 Nobel Prize in Physics.

Werner Heisenberg (1901–1976) – German physicist who formulated the Heisenberg uncertainty principle, which states that the more precisely the position of a particle is known, the less precisely its momentum can be determined, and vice versa, for which he won the 1932 Nobel Prize in Physics.

Peter Debye (1884–1966) – Dutch-American physicist and chemist who won the 1936 Nobel Prize in Chemistry for his work on dipole moments (the separation of charges in atomic bonds) and x-ray diffraction.

Lawrence Bragg (1890–1971) – British physicist who, along with his father, discovered Bragg's law of x-ray diffraction, a means of determining molecular structure, and shared the 1915 Nobel Prize in Physics at the age of 25, still the youngest-ever science winner.

Above: Lawrence Bragg, youngest-ever science Nobel laureate, who chaired later Solvay conferences.

Paul Dirac (1902–1984) – British physicist who formulated the Dirac equation, which predicts the existence of antimatter; he shared the 1933 Nobel Prize in Physics with Schrödinger, and famously wrote, "God used beautiful mathematics in creating the world."

Arthur Compton (1892–1962) – American physicist who demonstrated the particle nature of electromagnetic radiation, a radical idea at a time when the wave nature of light was widely accepted, for which he received the 1927 Nobel Prize in Physics.

Louis de Broglie (1892–1987) – French physicist who proposed that electrons – and, in fact, all matter – has wave properties, helping to lay the groundwork for wave–particle duality. He received the Nobel Prize in Physics in 1929.

Max Born (1882–1970) – German physicist and mathematician who helped to develop quantum mechanics and especially the statistical understanding of wave functions, for which he received the Nobel Prize in Physics in 1954.

Nils Bohr (1885–1962) – Danish physicist who developed the Bohr model of the atom, proposing that electrons occupy distinct orbits around the nucleus of an atom, and helped shaped the idea that wave and particle properties are complementary, receiving the Nobel Prize in Physics in 1922.

One of the most famous exchanges at the fifth conference took place between Einstein and Bohr. Disenchanted with Heisenberg's uncertainty principle, Einstein expressed his disbelief that God would play dice with

Left: 1922 Nobel Laureate in Physics Nils Bohr, who famously told Einstein to "stop telling God what to do".

the universe. Bohr famously responded with these words: "Einstein, stop telling God what to do!" Again, Marie was the only woman in attendance.

After the death of Lorentz, the next two Solvay conferences, in 1930 and 1933, were chaired by Langevin, another testament to the high regard in which he was held by other scientists. These were followed by five conferences between 1948 and 1961 chaired by Lawrence Bragg. The Solvay conferences continue today, and, befitting its founder's background, they are also held in chemistry, the first having taken place at Cambridge in 1922.

Above: William Henry Bragg, with his spectrometer, c. 1910s.

ALBERT EINSTEIN

PERHAPS THE BEST-KNOWN SCIENTIST OF THE TWENTIETH CENTURY WAS ALBERT EINSTEIN. BORN IN WHAT IS NOW GERMANY IN 1879, HE WAS THE SON OF AN ENGINEER WHO CO-FOUNDED AN ELECTRIC COMPANY.

Educated in Germany, Einstein excelled in mathematics and physics from a young age, and as a teenager sought admission to the Swiss polytechnic school in Zurich. Despite high marks in his strongest subjects, his overall score was too low to gain admission.

To avoid compulsory military service, Einstein renounced his citizenship, and soon fell in love with a fellow student at the polytechnic school, Mileva Marić. The two produced a daughter, but she was either given up for adoption or died early in life. They were married in 1903 and Mileva bore Einstein two sons, Hans Albert in 1904, and Eduard in 1910. The couple moved to Berlin in 1914 but were soon living apart. In 1919, Einstein married his cousin, Elsa, with whom he had been in a relationship since 1912.

As part of the divorce between Einstein and his first wife, which became official in 1919,

Below: Einstein as a boy of 14.

Right: Einstein at age 25.

Einstein agreed to hand over any money he might receive for winning the Nobel Prize. When he was awarded the 1921 Nobel Prize in Physics for his work on the photoelectric effect, he transferred the money to her. This support became particularly important in helping to pay medical expenses when Eduard was diagnosed with schizophrenia and eventually committed to an asylum.

In 1915, Einstein published a paper that contained what came to be known as his theory of general relativity. Somewhat analogous to the relationship he saw between mass and energy, he argued that gravity could curve space and time, which he referred to collectively as spacetime. This prediction has been verified by many observations, including the fact that massive objects such as the sun

Left: Einstein and his first wife, Mileva, in 1911.

Right: Einstein and his second wife, Elsa, in 1921.

THE "MIRACLE YEAR"

Early in his career, Einstein was unable to obtain an academic appointment and worked as a clerk in a Swiss patent office in Bern from 1902 to 1909. In 1905, the same year he received his PhD, he published four papers of such significance that it is often called his "miracle year". The first paper concerned the photoelectric effect. It argued that light energy can only be emitted or absorbed in discrete amounts, or quanta. This contradicted the view that light is strictly a wave phenomenon.

The second paper addressed Brownian motion, the seemingly chaotic movement of particles in a fluid. Einstein showed that such motion could be described statistically using the kinetic theory of gases, providing strong evidence for the existence of atoms and implying that they could be observed through a microscope. For the first time, atoms ceased to be hypothetical constructs and became observable phenomena.

The third paper articulated what came to be known as the theory of special relativity. Einstein suggested that the laws of classical mechanics no longer obtain at speeds approaching that of light, and connected such basic physical parameters as time, space, mass, and energy in a new way. It is based on the notion that the speed of light is fixed and does not depend on the position or state of motion of any particular observer.

The fourth paper includes perhaps the most famous equation in the world, $E = mc^2$. In essence, Einstein argued for the equivalence of matter and energy, suggesting that even completely stationary particles can be highly energetic. This equation specifies the amount of energy that can be released by nuclear reactions, such as the fusion reactions that power the sun and the fission reactions that make nuclear power plants possible.

bend light and alter the passage of time.

The rise of the Nazis in Germany led Einstein to emigrate to the United States in 1933, where he settled in Princeton, New Jersey, soon becoming an American citizen. On the eve of World War II, he warned the US not to fall behind the Germans in developing the atomic bomb but also campaigned against the weaponization of nuclear fission. By the time he died, he was the world's best-known scientist, a title that had once belonged to Marie Curie.

Einstein and Marie first met at the 1911 Solvay conference, when Einstein was 32 years old and Marie 43. When Einstein learned of the French press's assault on Marie, he was incredulous and leapt to her defence. Einstein branded the "horror story peddled in the newspapers" "nonsense". Yet he based this judgement less on his high regard for Marie's moral probity than on certain practicalities of the matter, as he saw them.

Specifically, Einstein's defence was grounded in his low estimation of her appeal to the opposite sex. He greatly admired Marie's intellect, work ethic, and commitment to scientific discovery. Of her passionate nature, he had no doubt. It was her physical charm he doubted. He wrote to a colleague that Marie Curie was "not attractive enough to represent a danger to anyone". She could not be a home-breaker, he opined, simply because no one would be moved to risk his family or reputation for her.

Left: Einstein receiving his US citizenship certificate in 1940.

Yet despite his doubts about the affair, Einstein offered strong words of defence and encouragement in a letter he wrote to Marie soon after the Solvay conference:

Do not laugh at me for writing to you without having anything sensible to say. But I am so enraged by the base manner in which the public is presently daring to concern itself with you that I absolutely must give vent to this feeling. However, I am convinced that you consistently despise this rabble, whether it obsequiously lavishes respect on you or whether it attempts to satiate its lust for sensationalism! I am impelled to tell you how much I have come to admire your intellect, your drive, and your honesty, and that I consider myself lucky to have made your personal acquaintance in Brussels. Anyone who does not number among these reptiles is certainly happy, now as before, that we have such personages among us as you, and Langevin, too, real people with whom one feels privileged to be in contact. If the rabble continues to occupy itself with you, then simply don't read that hogwash, but rather leave it to the reptile for whom it has been fabricated.

THE AFFAIR'S AFTERMATH

THE FINAL TWO MONTHS OF 1911 COMPRISED A MOMENTOUS PERIOD IN THE LIFE OF MARIE CURIE. JUST A DAY AFTER THE SOLVAY CONFERENCE ENDED, STORIES ALLEGING AN AFFAIR BETWEEN MARIE AND LANGEVIN BEGAN APPEARING IN FRENCH NEWSPAPERS.

Once she reached home, Marie sought legal advice and sent a letter to one of the newspapers, labelling the press's intrusion into her private life "abominable", and threatening legal action against papers that continued what she regarded as an assault on her privacy.

At the same time, word arrived that Marie had been awarded the 1911 Nobel Prize for Chemistry. She had not only played a crucial role in adding two new elements to the periodic table, but these elements opened the door to the study of radioactivity. The discovery had paved

Above: Marie Curie's diploma for the 1911 Nobel Prize in Chemistry.

the way for Rutherford's important work on the structure of atoms and opened a door to the use of radioactive elements such as radium in the treatment of diseases such as cancer.

Later in the month, a newspaper published excerpts from the correspondence between Marie and Langevin. These letters included many terms of endearment and references to their shared apartment, as well as advice to Langevin on how to end his marriage. These letters and subsequent editorials had the effect of building public sympathy for Madame Langevin and inciting ire against Marie, a "foreigner" whose actions were destroying a "French family".

In fact, the portrait of Marie sometimes verged on that of an evil genius:

> *This foreign woman, who pushes a hesitant father of a family to destroy his home … disposes of these poor people: of the husband, the wife, the children. … And she applies her scientist's subtlety in indicating the ingenious means by which one can torture this simple wife in order to make her desperate and to force the rupture.*

Angry crowds gathered outside Marie's house, scaring her daughters and effectively imprisoning her in her own home. Friends offered her the opportunity to stay in their spare rooms, away from the mob. Some urged Marie to get out of Paris, and others suggested that she leave the country altogether. Marie protested that she was French, that her daughters were French, and that they would not under any circumstances flee to Poland or anywhere else.

Above: King Gustav V of Sweden, reigning monarch and dinner companion to Marie Curie during the 1911 Nobel Prize ceremony.

LANGEVIN'S DUEL

Meanwhile, the publication of the private correspondence between Marie and Langevin convinced Langevin that he had no choice but to challenge the editor to a duel. The editor had not only written harshly of Marie but had also humiliated Langevin by calling him a "coward" who, though he had allowed his wife's reputation to be "dragged through the mud", had nonetheless appeared to be hiding behind her skirts.

At the appointed hour, Langevin and the editor arrived at a Paris park. The pistols were loaded. The two men were positioned at the proper distance. When the moment to fire finally arrived, however, the editor did not raise his pistol. Seeing this, Langevin lowered his. The duel was over, with no shot fired. The editor later claimed that, no matter how seriously Langevin had erred in his personal life, no Frenchman could deprive the nation of "such a precious brain".

Above: Duelling pistols.

Recognizing the problem that the scandal might present to the Nobel committee, Marie wrote to them, offering not to accept the 1911 Nobel Prize in Chemistry in person. Initially, the committee dismissed this suggestion as completely unnecessary, but when the Marie–Langevin correspondence was published and the committee learned of the duel, its members begged Marie not to attend the ceremony and even indicated that she might not have received the prize had the letters been published earlier.

This message precipitated a change of heart, and Marie wrote back that she would most certainly attend the ceremony, saying:

The action which you advise would appear to be a grave error on my part. ... I believe that there is no connection between my scientific work and the facts of private life. ... I cannot accept the idea in principle that the appreciation of the value of scientific work should be influenced by libel and slander concerning private life. I am convinced that this opinion is shared by many people. I am very saddened that you are not of this opinion.

In December, Marie attended the Nobel ceremony in Stockholm, accompanied by Irène and her sister Bronya. When she delivered her official lecture, she spoke of radioactivity as a parent would a child, describing the difficulty of her labours and how proud she was to behold its beauty. She remembered many others who had contributed to the "birth", including above all Pierre, and expressed the hope, subsequently realized, that radioactivity would permit the discovery of still more elements.

At virtually the same time, the legal proceedings between the Langevins were suspended: they would settle their dispute out of court; Langevin would admit that the fault was his; Madame Langevin would receive custody of the four children; and the father would get to see his children on Sundays and holidays. Moreover, Madame Langevin would receive a monthly payment from her husband. Marie had hoped the trial would go forward, but Langevin could not publicly oppose his children's mother.

Several years later, the Langevins achieved a reconciliation. Years after that, Langevin took another mistress, a young physicist who worked at the Curie laboratory in the Radium Institute and for whom Marie Curie later provided a successful recommendation for a scholarship. Her name was Eliane Montel (1898–1992), and in 1933 she bore Langevin a son, Paul-Gilbert Langevin (1933–1986), who went on to become an important musicologist.

There is no convincing evidence that Langevin and Marie were romantically

involved after 1911. Langevin spoke admiringly of her the rest of his life, and often expressed regret that he had not been able to do more in her defence. From Marie's side, relations remained cordial. But the end of their relationship had taken a great toll on her, and by the end of 1911 she was not only sick with grief but also physically quite ill.

Above: Paul-Gilbert Langevin, musicologist and son of Paul Langevin.

RECOVERY

ALTHOUGH MARIE AVIDLY FOLLOWED SCIENTIFIC DEVELOPMENTS OVER THE NEXT THREE YEARS, HER HEALTH PREVENTED HER FROM IMMERSING HERSELF IN THE LIFE OF A SCIENTIST. SHE WAS SUFFERING FROM PYELONEPHRITIS, AN INFECTION OF THE KIDNEYS.

Initially opposed to surgery, in March of 1912 she finally consented, but the nature of the operation is unknown. She was so sick afterwards that she made arrangements for the disposition of her property, should she die.

In 1913, Marie felt sufficiently well to travel to Warsaw to celebrate the opening of an institute dedicated to the study of radioactivity. There, for the first time in her life, she delivered a scientific lecture in her native language. She also saw first-hand the damage Russian rule was continuing to inflict on her motherland, leading her to lament its "absurd and barbarous domination" and marvelling at all that her countrymen were doing to "defend its moral and intellectual life".

Later in the year, Marie undertook a hiking expedition with Einstein in the Swiss Alps, accompanied by her daughters and Einstein's son. Ève recalls how, passing alongside crevasses and toiling up steep rocks, Einstein would suddenly stop dead in his tracks and expound some problem that was preoccupying him. Once he declared, "You understand, what I need to know is exactly what happens to the passengers in an elevator when it falls into emptiness."

Left: Albert Einstein and Marie Curie, c. 1929.

TRAVELLING RECUPERATION

Below: The Radium Institute in Paris, on which construction began in 1911.

Over the course of her long recovery, Marie travelled from place to place. In each new locale she hoped to find living conditions conducive to her recovery, but she was also avoiding public scrutiny. She used different pseudonyms and kept her address a secret, to avoid detection by the press. Her itinerant mode of life meant suspending her research and teaching, for which she begged the understanding of her colleagues at the Sorbonne.

Marie applied her scientific mind to her own recovery, recording precisely the amount of water she took in and the amount of urine she produced every day. In addition, she noted whether the urine contained pus, her temperature, and the location and character of the pains she was experiencing. There is little evidence that such meticulous observations contributed to her recovery, but they at least provided her with a means of coping with an illness over which she seemed to exercise little influence.

Back in Paris, the University of Paris and the Pasteur Institute were joining forces, at a site on a street named after Pierre, to build a Radium Institute, which would be directed by Marie. The institute would conduct both physical research into radioactivity and biomedical investigations into its use in the treatment of cancer and other diseases. While the buildings were going up, Marie tended to the garden between them, even planting many of the flowers herself.

Ève reports that when the new Radium Institute and its Curie pavilion were finally opened in the summer, Marie evoked the words of Pasteur:

> *If conquests useful to humanity touch your heart, if you stand amazed before … so many admirable discoveries, if you are jealous of the part your country can claim in the further flowering of these wonders – take an interest, I urge upon you, in those holy dwellings to which the expressive name of laboratories is given. Ask that they be multiplied and adorned. They are the temples of the future, of wealth and well-being. It is there that humanity grows bigger, strengthens and betters itself. It learns there to read in the works of nature, works of progress and universal harmony, whereas its own works are too often those of fanaticism, barbarity, and destruction.*

Sadly, before the year 1914 had drawn to a close, Pasteur's dreams of inexorable progress and universal harmony, which Marie certainly shared, were dashed to pieces.

Left: Louis Pasteur, one of the most famous of all French scientists, whose words Marie Curie invoked at the opening of the Radium Institute.

Above: The Curie Museum resides within the walls of the original Radium Institute.

WAR

THE UNPRECEDENTED DESTRUCTION WREAKED BY WORLD WAR I – INCLUDING THE DEATHS OF MORE THAN 16 MILLION SOLDIERS AND CIVILIANS – CAN BE TRACED TO THE ASSASSINATION OF ARCHDUKE FRANZ FERDINAND IN SARAJEVO IN JUNE OF 1914.

Above: Austro-Hungarian Archduke Franz Ferdinand with his wife Sophie and their three children.

In this act of rebellion against Austro-Hungarian rule over Bosnia and Herzegovina, Austria-Hungary saw an opportunity to suppress nationalists. With assurances from Kaiser Wilhelm II that Germany would support the effort, Austria-Hungary issued an all-but-impossible ultimatum.

Serbia's troops mobilized and appealed to Russia for assistance. Within a matter of days, the battle lines were drawn, with Belgium, England, and France allied against Austria-Hungary and Germany. Germans adopted a two-front strategy, engaging the Russians in the east and the Belgians in the west. Having quickly overwhelmed the Belgians, the Germans commenced their advance through France toward Paris.

In September, the British and French forces confronted the Germans at the First Battle of the Marne, just 30 miles outside Paris. Over several days of intense fighting, the Allied forces were able to turn back the Germans, saving Paris. Over the next year, additional major battles would be fought, including the Battle of the Somme and the Battle of Verdun, each of which produced approximately 1 million casualties.

The scale of carnage in World War I vastly exceeded previous conflicts, in part because of innovations in the technology of war. Tanks, flamethrowers, and poison gases were utilized on a large scale for the first time. In addition, the placement of troops in military

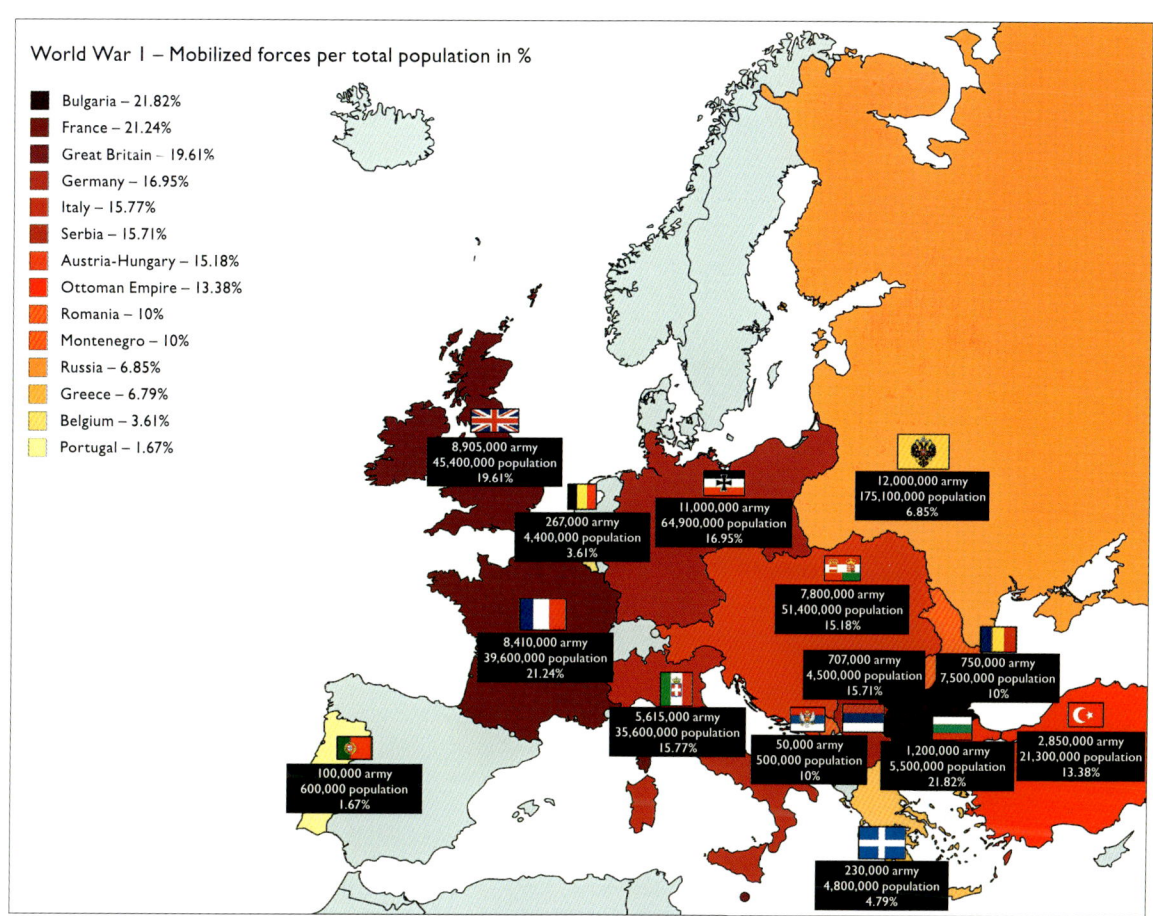

World War I – Mobilized forces per total population in %

- Bulgaria – 21.82%
- France – 21.24%
- Great Britain – 19.61%
- Germany – 16.95%
- Italy – 15.77%
- Serbia – 15.71%
- Austria-Hungary – 15.18%
- Ottoman Empire – 13.38%
- Romania – 10%
- Montenegro – 10%
- Russia – 6.85%
- Greece – 6.79%
- Belgium – 3.61%
- Portugal – 1.67%

8,905,000 army
45,400,000 population
19.61%

11,000,000 army
64,900,000 population
16.95%

12,000,000 army
175,100,000 population
6.85%

267,000 army
4,400,000 population
3.61%

7,800,000 army
51,400,000 population
15.18%

8,410,000 army
39,600,000 population
21.24%

707,000 army
4,500,000 population
15.71%

750,000 army
7,500,000 population
10%

5,615,000 army
35,600,000 population
15.77%

50,000 army
500,000 population
10%

1,200,000 army
5,500,000 population
21.82%

2,850,000 army
21,300,000 population
13.38%

100,000 army
600,000 population
1.67%

230,000 army
4,800,000 population
4.79%

trenches, so-called "trench warfare", made it difficult for either side to advance – one of the reasons poisonous gases were introduced. Trench warfare soon became synonymous with stalemate and futility.

On the eastern front, Russia's efforts to break the German lines were unsuccessful, and domestic economic difficulties and military stalemate produced rising hostility toward Czar Nicholas II's regime. These tensions culminated in the Russian revolution of 1917 and the rise to power of Vladimir Lenin (1870–1924)

and the Bolsheviks. When Russia signed an armistice with the Central Powers (Germany, Austria-Hungary, the Ottoman Empire, and Bulgaria) in 1917, the Germans seized the opportunity to turn all their attention to the western front.

Later in 1917, driven in part by the German sinking of multiple passenger and merchant ships, including the *Lusitania*, the Americans entered the war on the Allied side. As German forces shifted to the western front, a Second Battle of the Marne was fought against both

Above: Map showing percentage of forces mobilized by nation in World War I.

French and American troops. When the Germans were again repulsed, followed days later by an Allied counteroffensive, Germany's hopes of victory began fading fast.

The war wound to a close in November of 1918, and a peace conference in 1919 ended with the Treaty of Versailles. Designed to ensure that the Great War would be the "war to end all wars", the treaty in fact produced the opposite effect, imposing heavy war reparations on Germany and preventing its entry into the League of Nations. By the 1930s, German resentment over the perceived humiliation and unfairness of the treaty helped sow the seeds of World War II.

Left: Czar Nicholas II of Russia.

Below: The British ship *Lusitania*, which was sunk off the coast of Ireland in May 1915.

POLAND REBORN

For Marie Curie, however, the horror of war did offer one recompense. Poland, which had existed under partition between Austria-Hungary, Germany, and Russia, would be reborn, thanks both to the US entry into the war and the Russian revolution. US President Woodrow Wilson (1856–1924) saw the war's end as an opportunity to spread democracy throughout Central Europe, in part by liberating the Poles from the rule of the Central Powers. The Second Polish Republic was at last established in 1918.

Below: Map of Europe after World War I, showing the restoration of Poland

MILITARY RADIOLOGY

WITH THE ONSET OF WAR AND THE MOBILIZATION OF TROOPS IN THE SUMMER OF 1914, MARIE FOUND HERSELF IN A RATHER LONELY SITUATION.

Nearly all her colleagues had gone off to war, and her daughters were on vacation at the seaside. Marie had planned to join them but felt she could not leave until the laboratory was in order. Soon she realized that the mobilization of troops might preclude her travel. Remaining in Paris, she urged the girls, particularly Irène, to be patient and brave.

Sending for the girls was out of the question. The advance of German troops meant that Paris might soon be occupied. Yet Marie also felt that she could not leave. The Germans might destroy her laboratory, or worse yet, confiscate its precious radium. To ensure the safety of the radium, Marie soon arranged to transport it to a more secure location in Bordeaux, which she would accomplish herself with the precious cargo sealed in a heavy lead case.

Marie was fully committed to her adopted country, the native land of her husband and daughters. She had left the large sum of money she received for her second Nobel Prize in a Swiss bank account, but to assist the war effort she purchased French war bonds, knowing full well that most or all the money would likely be lost. As she suspected, this is exactly what happened, although her generosity did buy her the assurance that she was doing all she could.

Marie also made plans to have her family's gold Nobel Prize medals melted down, so that the proceeds could be used to support the war effort. However, officials at the bank, recognizing the national significance of the medals, refused. If Marie wished to provide further aid to France, she would need to do so not by donating her treasure, but instead through the investment of her own time and prodigious talent.

Fortunately, such an opportunity presented itself. X-rays, which had been discovered by Roentgen just two decades earlier, made it possible to peer into the living human body without cutting it open. In wartime, this ability could enable physicians to locate and remove bullets and shrapnel from the bodies of the wounded without the need for often-extensive exploratory surgery. This in turn would lower the risk of infections, amputations, and death.

As France's greatest physicist, Marie foresaw the military potential of x-rays better than anyone. She made enquiries to determine whether the France army possessed

Left: Marie Curie driving one of the vehicles she outfitted as a mobile radiology facility.

Below: One of the mobile radiology units used by the French military during World War I.

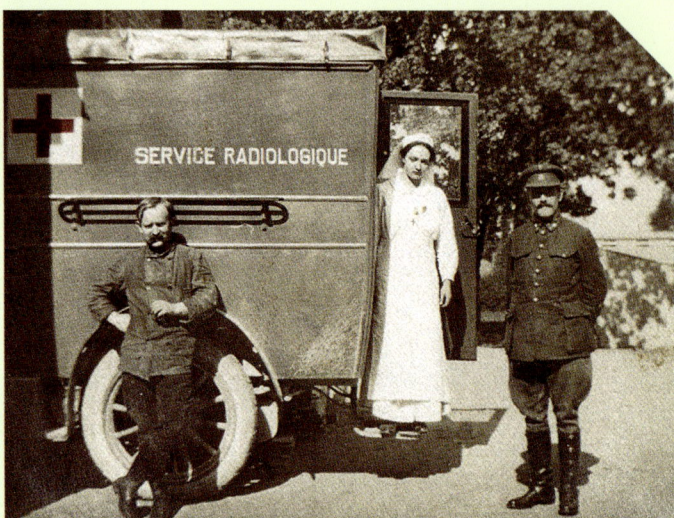

such equipment, the personnel to operate it, and the know-how to maintain it. Quickly satisfied that this was not the case, a path opened before her. She could make very practical use of her scientific and technical expertise, aid her nation at a critical juncture in its history, and render humanitarian aid to wounded soldiers and civilians.

Marie saw that the contributions of the x-ray to the war effort would be limited so long as x-ray facilities were confined to hospitals. Occasionally, battles might take place in the vicinity of such a facility, but in most cases casualties occurred some distance away. To address this problem, Marie set to work devising mobile x-ray units. While the machines themselves were not heavy, the principal problem would be to provide sufficient electric current for their operation.

Marie's solution was ingenious. She would cajole friends and acquaintances into loaning her their vehicles. Then she would outfit them with x-ray machines, examination tables, and other equipment. At first, she planned to install gas-powered electrical generators in the vehicles as well, but she soon realized that this would add hundreds of pounds of additional weight. Instead she would use the vehicle's engine to power the generator.

Seeing her first radiology car in action, Marie quickly became convinced that many more were needed. She lobbied hard to obtain vehicles, x-ray equipment, and personnel dedicated to the cause. Not only did she eventually outfit approximately 20 such cars herself, but she also helped to set up hundreds of stationary radiology facilities in the field. Some of her facilities are estimated to have contributed to the care of more than 10,000 casualties.

Of course, the challenges were not purely technical. She also had to convince physicians, many of whom had little experience with x-rays, of the utility of the new imaging technique, teaching them how to use radiographs to locate precisely foreign bodies in the tissues. She also had to convince a sceptical military bureaucracy, which, despite Marie's fame, dedication, and evident organizational skills, was disinclined to admit that a woman had any business racing back and forth behind battle lines.

Despite her prodigious imagination and energy, Marie could not fail to recognize her own limitations. Even if she had worked 24 hours a day without respite, one person could contribute only so much. So she turned a portion of her attention to educating others. Specifically, she began training technicians – mainly women – to operate x-ray machines. She established a school for this purpose, which attracted women from many different stations in life, eventually graduating 150.

LIKE MOTHER, LIKE DAUGHTER

Irène, desperate to make herself useful, eventually joined her mother at the school. Having initially come to Paris in the autumn, she set aside her university studies and enrolled in nursing school. She then joined her mother on the road, learning everything she could about the new radiological equipment. Still just a teenager, she was soon following in her mother's footsteps, travelling from facility to facility and helping to solve a wide variety of technical problems.

While working full-time as a teacher at the school for radiological technicians, Irène also found the energy to resume her university studies, obtaining qualifications in maths, physics, and chemistry with distinction – just the path her father and mother had followed. It brought Marie great pride to see her daughter labouring so tirelessly and with such a high level of excellence in service to her country and humanity.

Right: Marie and Irène Curie at a Belgian hospital in 1915.

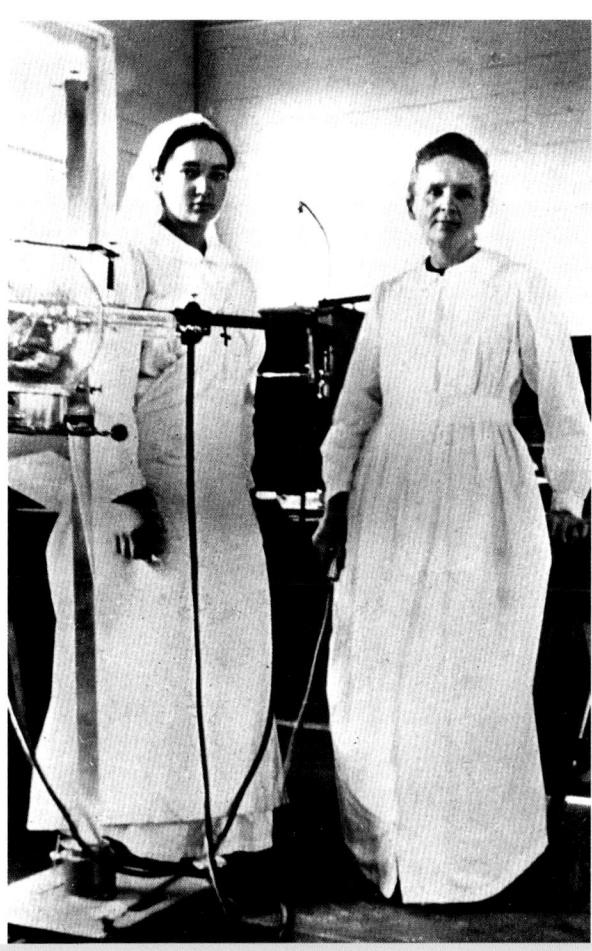

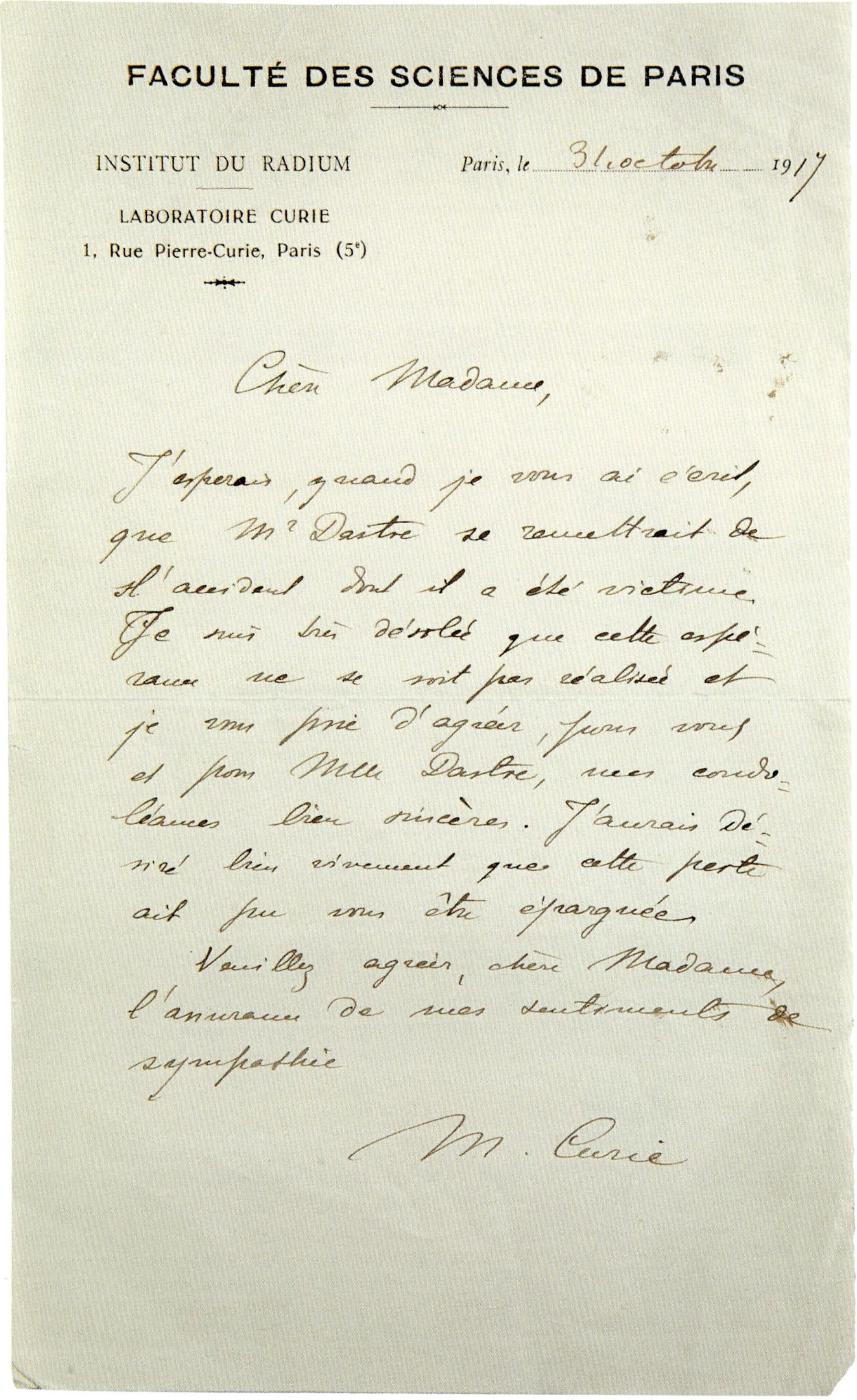

FACULTÉ DES SCIENCES DE PARIS

INSTITUT DU RADIUM

Paris, le 31 octobre 1917

LABORATOIRE CURIE

1, Rue Pierre-Curie, Paris (5ᵉ)

Chère Madame,

J'espérais, quand je vous ai écrit,
que Mr Dastre se remettrait de
l'accident dont il a été victime.
Je suis très désolée que cette espé-
rance ne se soit pas réalisée et
je vous prie d'agréer, pour vous
et pour Mlle Dastre, mes condo-
léances bien sincères. J'aurais dé-
siré bien vivement que cette perte
ait pu vous être épargnée

Veuillez agréer, chère Madame,
l'assurance de mes sentiments de
sympathie

M. Curie

Over the course of the war, Ève Curie estimates that the radiological facilities her mother spearheaded helped to care for over one million wounded. Marie eventually became convinced that a driver was a luxury she could not afford, so she obtained her own driver's licence. On more than one occasion, this two-time Nobel laureate could be seen under the hood, cleaning a carburettor or hunched down beside a vehicle, changing a tyre.

Preparing to venture forth one day in 1915, Marie wrote to Langevin:

> *The day I leave is not fixed yet, but it can't be far off. I have received a letter saying that the radiological car working in the Saint-Pol region has been damaged. This means that the whole north is without any radiological service! I am taking the necessary steps to hasten my departure. I am resolved to put all my strength at the service of my adopted country, since I cannot do anything for my unfortunate native country just now, bathed as it is in blood after more than a century of suffering.*

For stretches of many months, Marie heard nothing from her native country or her family and friends there. Were they alive? What hardships was the war inflicting on them? This made the war's end all the sweeter, and when armistice finally came, Marie and some of her friends outfitted one of the radiology cars with a homemade French flag and drove up and down the streets of Paris, celebrating the end of hostilities.

Once communication with Poland was re-established and her mother country had been reconstituted as an independent nation, Marie wrote to her brother Joseph:

> *So now we, "born in servitude and chained since birth," have seen this resurrection of our country which has been our dream. We did not even hope to live to this moment ourselves; we thought it might not even be given to our children to see it – and here it is! It is true that our country has paid dearly for this happiness, and it will have to pay again. But can the clouds of the present situation be compared with the bitterness and discouragement that would have crushed us if, after the war, Poland had remained in chains and divided into pieces? Like you, I have faith in the future.*

To her surprise, Marie would soon find herself travelling to one of France's most important wartime allies, the United States. The woman who would engineer this journey and later become an important friend of Marie's was a newspaper editor named Missy Meloney.

Left: Handwritten letter from Marie, dated 31 October 1917.

MISSY MELONEY

MARIE MELONEY, KNOWN TO MOST AS "MISSY", WAS BORN IN KENTUCKY IN 1878 TO A PHYSICIAN FATHER AND A MOTHER WHO FOUNDED THE *KENTUCKY MAGAZINE*.

Initially trained as a pianist, a horse-riding accident led her to follow her mother into journalism. She also suffered from respiratory infections. A feminist, she later remarked, "I have been lame since 15, had a bad lung since 16, and have done the work of three men ever since." In 1904 she married a New York editor and later become secretary to the city's mayor.

Above: Cover of an 1894 issue of *The Delineator*, which was edited by Missy Meloney from 1920.

By 1920, Meloney had become editor of *The Delineator*, a women's magazine. For some years, she had been attempting to arrange an interview with Marie Curie by one of her writers, but Marie rebuffed such advances. Finally, Meloney travelled to Paris herself, sending a note to Marie in which she indicated that she had been trying for two decades to arrange an interview. Marie, perhaps impressed by her persistence, granted her request.

Meloney described her first impression of Marie:

> The door opened and I saw a pale, timid little woman in a black cotton dress, with the saddest face I ever looked upon. Her kind, patient, beautiful face had the detached expression of a scholar. Suddenly I felt like an intruder. My timidity exceeded her own. I had been a trained interrogator for twenty years, but I could not ask a single question of this gentle woman. … I tried to explain that American women were interested in her great work, and found myself apologizing for intruding upon her precious time.

Marie expressed to Meloney that the one thing she wanted most was a gram of radium so she could continue her research. Though leading a fairly comfortable life at this point, post-war economic circumstances in France

Left: Missy Meloney
and Marie Curie
in 1921.

left her without the means to acquire the substance, the price of which hovered around $100,000 per gram. Meloney, moved by the thought that radium's discoverer was bereft of the substance, pledged to raise the money from American women.

Meloney offered just one condition. If she could raise the money, primarily through small donations, Marie must agree to travel to the United States to accept the gift. Marie agreed, but before doing so, insisted that Meloney secure the agreement of newspaper editors not to publish any stories of the affair with Langevin. Marie, knowing that more than 50 grams of radium could already be found in American labs, hoped that she had found a way to continue her research.

When Meloney died in 1943, having most recently edited *This Week* magazine for a decade, Eleanor Roosevelt wrote of her:

One never came away depressed from seeing "Missy" Meloney. If I am sometimes weary and think that perhaps there is no use in fighting for things in which I believe against overwhelming opposition, the thought of what she would say will keep me from being a slacker. She believed that women had an important part to play in the future. She not only helped such women as Marie Curie, who were great women, but she helped many little people like myself to feel that we had a contribution and an obligation to try to grow.

Left: Missy Meloney with Irène, Marie, and Ève Curie in 1921.

A VARIED CAREER

Throughout her life, Meloney had other claims to fame. As a teenager, she had been one of the first female reporters to obtain a seat in the US Senate press gallery. In the 1920s, she organized a Better Homes in America campaign, with future US Vice President Herbert Hoover (1874–1964) at the head of the organization. She organized national conferences on the health of women and children. And she was a close friend and confidante of First Lady Eleanor Roosevelt (1884–1962).

Right: Eleanor Roosevelt, who eulogized Missy Meloney at the United Nations in 1947.

MARIE IN AMERICA

WHEN MARIE AND HER DAUGHTERS SAILED FOR THE UNITED STATES IN THE SPRING OF 1921 TO ACCEPT THE GIFT OF A GRAM OF RADIUM, THEY DID SO UNDER SOMEWHAT FALSE PRETENCES THAT HAD BEEN CRAFTED BY MISSY MELONEY.

In her appeals to American women to assist Marie's important scientific work, Meloney took some liberties in her portrayal of Marie's material circumstances, painting them as considerably more straitened than an objective account would bear.

It was important to Meloney that Marie should visit the US in the spring, so that colleges and universities would still be in

MARIE AS A SYMBOL

In Meloney's hands, Marie became the symbol of a suffering post-war France. Although she had her own laboratory, lived in a spacious apartment, and drew both a professor's salary and a modest pension from the French government, Meloney missed no opportunity to emphasize the magnitude of Marie's struggles. This portrayal of Marie in the American press as impoverished engendered resentment among some of her fellow countrymen.

Another, and from Marie's point of view more troubling, aspect of Meloney's public relations campaign was her decision to portray radium and Marie's work with it as the cure for cancer. While radium held medical promise, no well-informed physicians or scientists were promoting it as the cure for cancer. Moreover, Marie had no intention of using the radium she received for medical purposes. She intended to continue her chemical research.

session and Marie would be available to receive honorary degrees. Marie refused to play any part in fund raising, though, in addition to an extensive travel schedule, she agreed to visit such natural wonders as Niagara Falls and the Grand Canyon. Marie's visit would stretch to seven weeks, during which she would be feted at dozens of dinners and receive ten honorary degrees and many awards.

When the ship arrived in New York, Marie's party was greeted by a large international press corps and bands performing both the French and US national anthems. After a tour of east-coast colleges, Marie was the guest of honour at a meeting of the American Association of University Women, which contributed to her impression that in America women are encouraged to pursue their aspirations. She also noted the stark contrast between dour French students and the smiling and excited faces of the Americans.

Opposite: Announcement in *Science* of Marie Curie's visit to America in 1921.

Above: Marie Curie during her tour of the United States in 1921 – here with Dean Pegram of the School of Engineering at Columbia University.

Right: Missy Meloney (second from left), Marie Curie (fourth from left), US President Warren Harding, and Irène Curie (sixth from left) during the 1921 tour.

Marie attended a reception hosted by US President Warren Harding (1865–1923) at the White House. It was, as Marie later recalled, with pun perhaps intended, "a radiant day in May". After a speech praising the friendship between France and the US, Harding presented her with the key to a case said to contain a gram of radium. In fact, of course, the radium was kept in an industrial safe, and the key and case were just for show.

Thereafter, Marie was exhausted. At many later stops, Irène and Ève attended dinners, spoke with the press, and accepted degrees on their mother's behalf. Physicians who were consulted during the visit blamed Marie's ill health on overwork, but she suspected that her work with radium had damaged her health. On her way to the Grand Canyon, Marie's entourage stopped in Chicago, where she especially enjoyed visiting the city's large Polish community.

When Marie sailed back to France, she left with a lead case containing ten hermetically sealed glass tubes, each containing approximately 100 mg of radium. The container is now on display in the Curie Museum in Paris. Before leaving, Marie had toured the radium refining plant in Pittsburgh, where the radium had been prepared. Though exhausting, the trip had provided Marie with what she longed for most: radium with which to continue her research.

Marie returned to America once again in 1929, again to receive a gift of radium from the American people. This time, however, Marie would use the radium not for her research in Paris, but to assist the new Radium Institute in Warsaw. And she did not receive the radium itself but a check for $50,000,

Opposite: Marie Curie and US President Warren Harding descending the White House steps in 1921.

Below: Certificate for the radioactive material delivered to Marie Curie during her 1921 US visit.

the cheapest price of a gram of radium at the time. This second visit received less attention, largely because it coincided with the great stock market crash that heralded the onset of the Great Depression.

IRÈNE AND FRÉDÉRIC

FRÉDÉRIC JOLIOT WAS BORN IN 1900 IN PARIS TO A LARGE, PROSPEROUS FAMILY. THE DEATH OF HIS FATHER MADE IT IMPOSSIBLE FOR HIM TO STUDY AT A PRESTIGIOUS UNIVERSITY, SO HE ATTENDED THE TECHNICAL UNIVERSITY IN WHICH BOTH PIERRE AND LANGEVIN HAD TAUGHT.

After Frédéric completed compulsory military service, Langevin suggested to Marie that he work in her laboratory at the Radium Institute. Marie was not initially impressed, but she placed Frédéric under the supervision of Irène.

Frédéric, or Fred, as he was known, presented a rather stark contrast to Irène. Irène could be rather preoccupied, often forgetting to say hello to her co-workers when she arrived at the Institute. She was also "all business" at work. Frédéric, on the other hand, was a gifted athlete and highly gregarious, the sort of person who quickly endeared himself to everyone he worked with. Though three years Irène's junior, he was also quite dashing.

Below: Irène Curie accepting an honorary degree at the University of Pittsburgh on behalf of her mother in 1921.

In 1925, just a year after Frédéric joined the lab, Irène presented her doctoral thesis, which she had prepared under the supervision of Langevin. The event was attended by a large crowd, including domestic and overseas reporters. Blessed with a larger supply of polonium than any other lab in the world, Irène had studied its emission of alpha rays. Asked how a woman could conduct such groundbreaking research, she calmly replied that the abilities of women were no different from those of men.

Each day at the lab, Frédéric would remain after work to walk Irène home, and soon the two were taking long walks together in the country. Looking back, Frédéric later described their blossoming relationship in these terms:

> *In observing her, I discovered that this young woman, whom others saw as a bit brutish, an extraordinary poetic and sensitive being who, in a number of ways, was a living representation of her father. I had read a lot about Pierre Curie, and I found in his daughter the same purity, good sense, and tranquillity.*

Above: Irène and Marie Curie in the laboratory in 1925.

Left: Irène and Frédéric Joliot-Curie in the laboratory.

Because Irène travelled abroad with Marie on several occasions, often setting up and conducting the experimental demonstrations for her talks, Irène had occasion to write to Fred, and these letters reveal the true depth of her feeling. She wrote of her longing to embrace him upon her return. For his part, he told her that since her departure, the lab seemed empty. Wed in a civil ceremony in 1926, they both changed their surnames to Joliot-Curie.

Irène became pregnant soon thereafter, and Marie's first grandchild, Hélène, was born the following year. Shortly after Hélène's birth, Marie attended the fifth Solvay conference, which produced such sparks between Einstein and Bohr. With his mother-in-law's urging, Fred went on to pursue both a baccalaureate and doctorate, producing a thesis on radiochemistry. In 1932, Irène and Fred welcomed a son, Pierre. Years later, Hélène became a nuclear physicist and Pierre a biochemist.

In 1935, the Joliot-Curies shared the Nobel Prize for Chemistry. Their feat ushered in a new era of relatively inexpensive radioisotopes, which quickly began to play an important role in both biological research and medical treatment. Irène and Frédéric both became professors in the faculties of science at the College of France in 1937. Frédéric built a cyclotron for the production of radio-isotopes and conducted research on the fission of uranium.

Below: Irène and Frédéric Joliot-Curie at the 1935 Nobel Prize ceremony.

GROUNDBREAKING RESEARCH

By the 1930s, the Joliot-Curies' experiments had produced some surprising results. They showed that when beryllium was bombarded by alpha particles, it produced radiation of greater power than the bombardment. A young colleague of Ernest Rutherford, James Chadwick, was the first to correctly interpret the results: the Joliot-Curies were observing a particle previously postulated by Rutherford – the neutron. The Joliot-Curies had revealed something new but failed to interpret their results correctly.

Soon, however, they conducted an experiment that introduced a new epoch in twentieth-century science. Bombarding aluminium with alpha particles, they discovered that they were producing "artificial radioactivity", creating short-lived radioisotopes and in effect transmuting one element into another, thereby realizing the dream that had long haunted the forerunners of chemistry, the alchemists. They were turning aluminium into a radioisotope of phosphorus, which then decayed into silicon.

With the onset of World War II, the Joliot-Curies placed their research papers on nuclear fission in a vault, and as the Germans approached in 1940, Frédéric transported his materials to England. In 1946, Irène became director of the Radium Institute, and in 1948, the Joliot-Curies helped to build the first French nuclear reactor. Thanks in large part to their legacy, France went on to derive 75 per cent of its electricity from nuclear power.

In the 1950s, Irène's health began to fail. She was suffering from fevers and weight loss, and she was diagnosed with leukaemia. She died of the disease in March of 1956 at the age of 58. Frédéric assumed her chair at the university and became head of the Curie Institute, but he too died just two years later of liver disease. In both cases, physicians speculated that exposure to radioisotopes may have precipitated their deaths.

In Irène's obituary, James Chadwick wrote of her:

She knew her mind and spoke it, sometimes perhaps with devastating frankness; but her remarks were informed with such regard for scientific truth and with such conspicuous sincerity that they commanded the greatest respect in all circumstances. In all her work, whether in the laboratory, in discussion, or in committee, she set herself the highest standards and she was most conscientious in the fulfillment of any duties she undertook.

Below: Marie, Irène, the Joliot-Curie children (Hélène and Pierre), Frédéric, and Frédéric's mother Emilie.

DEATH OF MARIE

MARIE CURIE'S LAST YEARS WERE DIFFICULT. HER FINGERS, LIKE PIERRE'S YEARS BEFORE, HAD BECOME STIFFENED, PRESUMABLY DUE TO RADIATION EXPOSURE.

Her vision was clouded by cataracts, an additional, now-predictable consequence. Her capacity for work was diminished, but this increased her satisfaction at knowing that Irène and Frédéric, about whom she had initially harboured some doubts, were now carrying on her scientific work.

Frédéric thought that Marie's last great satisfaction in life came from witnessing the transmutation of elements that he and Irène had accomplished. Recalling how he repeated the experiment for Marie and Langevin late in the afternoon on that fateful 1934 day, he wrote:

> I will never forget the expression of intense joy which came over Marie when Irène and I showed her the first artificially radioactive elements in a little glass tube. I can still see her taking in her fingers (which were burnt with radium) this little tube containing the radioactive compound – in which the activity was still very weak. To verify what we had told her, she held it near a Geiger-Müller counter and she could hear the rate meter giving off a great many clicks. This was doubtless the last great satisfaction of her life.

Marie's visits to the lab became less and less frequent. She left early on the last day in May of 1934, complaining of a fever and headache. The physician who examined her obtained an x-ray image of her chest and diagnosed tuberculosis. Arrangements were made for her

Right: Marie Curie in 1931.

to travel to a sanatorium in the French Alps, accompanied by Ève. The difficult journey complete, further tests were performed, which argued against the tuberculosis diagnosis. Instead, the physicians said, she had severe anaemia.

Ève recalled her mother's last hours:

On the morning of July third, for the last time Madame Curie could read the thermometer held in her shaking hand and distinguish the fall in fever which always precedes the end. She smiled with joy. This was the sign of her cure, and she was going to be well now, she said. Looking at the open window, turning hopefully toward the sun and the motionless mountains, she said, *"It wasn't the medicines that made me better. It was the pure air, the altitude."*

Marie died early the next morning, 4 July 1934. In his final note, her physician noted, "The disease was an aplastic pernicious anaemia of rapid, feverish development. The bone marrow did not react, probably because it had been injured by a long accumulation of radiations." Subsequent physicians would speculate that Marie had died of a form of leukaemia which had prevented the normal progenitor cells in her bone marrow from making new red and white blood cells.

On this same day, Leo Szilard (1898–1964), a Hungarian physicist, filed a patent on a device that could produce an immense

Above: Leo Szilard, who, on the day Marie
Curie died, filed a patient on a design for a
nuclear bomb.

Left: Marie Curie at a Cancer Society dinner
in 1929.

Below: Tomb of Marie and Pierre Curie in the Panthéon in Paris.

explosion relying on the energy contained in the atomic nucleus. Years later, in the United States, Szilard participated in conducting the first sustained nuclear chain reaction, which would soon lead to the development of the so-called "atomic bomb". Two such bombs, dubbed "Little Boy" and "Fat Man", were subsequently dropped on the cities of Hiroshima and Nagasaki near the end of World War II.

Marie was buried on 6 July in a small ceremony attended only by family and a few close friends. There was no pomp and no flowery speeches. Instead the ceremony was kept very short and simple, just as Marie and Pierre would have wanted it. Her coffin was lowered onto Pierre's and sprinkled with Polish soil brought by her siblings from their motherland. Although photographs showed a woman who looked a decade older than her 66 years, her brilliant life had burned out prematurely.

In 1995, the remains of Pierre and Marie Curie were exhumed and reinterred in the Panthéon in Paris, the resting place of France's greatest heroes. Marie became the first woman so honoured based on her own merits, joining such French cultural giants as Voltaire (1694–1778), Rousseau (1712–1778), and Victor Hugo (1802–1885). Already entombed in the Pantheon were two of Marie's colleagues in physics, Jean Perrin (1870–1942) and Marie's partner and long-time friend, Paul Langevin.

At the time of the exhumation, radiation measurements were obtained of the coffins and the remains of Marie and Pierre. Would the measurements disclose high levels of radioactivity, evidence that the Curies had been killed by the radioisotopes they discovered? The joint statement of the two teams that performed the measurements reads as follows:

The exhumation of Pierre and Marie Curie was conducted Friday, 14 April, 1995 and lasted 1 hour and 30 minutes. Dose rates measurements, air sampling, and analysis of the wood coffins did not show the presence of significant radioactivity. The results confirmed the absence of radiological risk for workers, the public, and the environment. As might be expected, traces of radium-226 have been detected at the coffins of Pierre and Marie Curie.

Experts who reviewed the case have suggested that, because of the low levels of radiation found in Marie's body, her death was probably less attributable to contamination by radium or polonium than the many x-ray examinations she helped to perform during World War I. That little radioactivity remained in her body at the time of her death does not prove that she was never contaminated with radium, however, because radium is gradually eliminated from the living body over a period of years.

One thing is certain. The Curies' laboratory equipment and notebooks from the 1890s are radioactive, so much so that they are considered too dangerous to handle without appropriate protection. In fact, even Marie's cookbook is radioactive. Such artefacts are kept shielded, and those who wish to inspect them directly must don protective clothing and sign a waiver. The 1,600-year half-life of radium means that such measures will remain necessary well into the future.

Above: Article detailing Marie Curie's death, *The Daily Mirror*, 5 July 1934.

MARIE'S LEGACY

MARIE CURIE WAS A REMARKABLE HUMAN BEING. BLESSED WITH UNDOUBTED INTELLIGENCE, SHE ALSO MANIFESTED EXTRAORDINARY CURIOSITY AND SINGLENESS OF PURPOSE.

Above: Plaque commemorating the first Maria Curie-Skłodowska laboratory, Warsaw, Poland.

She could persevere in circumstances that would lead almost anyone else to admit failure, and many of the setbacks she experienced only augmented her resolve. She knew why she had been placed on this earth, and she did her level best to devote herself to that mission every day.

She was also a deeply loyal person. Born into a country that didn't exist, she remained staunchly devoted to Poland and contributed as much as anyone to the pride of its people. Having adopted the country of her husband and daughters, she was also fiercely devoted to France, even changing her name from Maria to Marie. Despite attacks in the French press, she displayed great generosity, ingenuity, and dedication in service to France during the war and afterwards.

And yet, even though Marie ranks as one of the greatest figures in the history of science, she was also cut from the same human cloth as the rest of us. She was capable of deep love, as expressed above all in her relationship with Pierre. They seem to have been an almost perfectly matched couple, different in all sorts of ways but also deeply complementary to one another and united by their dream of a life consecrated to scientific research.

She suffered deeply from many of the losses to which love exposes the human heart. Wounded as a child by the deaths of her sister and mother, she found herself deprived her of the capacity to believe in a benevolent god. She was plunged into despair at the loss of her first love, the end of her relationship with Langevin, and especially by the death of Pierre. Despite such losses, she was able to summon within herself the will to carry on.

A LEGACY OF FIRSTS

Marie fought and overcame deeply entrenched bias. Despite being poor, Polish, and a woman, she managed to get herself a first-rate education, gain access to research facilities and mentors, and win a host of unprecedented recognitions: the first female professor at the University of Paris, first woman to win a Nobel Prize, only woman to win two Nobel Prizes, and the only human being ever to win Nobel Prizes in two different natural sciences.

Above: 1937 photo of Ève Curie, who wrote the best-known biography of her mother Marie, *Madame Curie*. She died in New York at the age of 102.

Right: Polish bank
note bearing the
image of Marie Curie.

She despised many of the things we are taught to prize. She had no interest in power in the conventional sense and worked through organizations only to protect and promote scientific pursuits. Wealth was a matter of indifference, except insofar as it served those same scientific purposes. Fame meant so little to her that she often insisted that prizes and honours go not to her but to her institution. Einstein would later write that she was the only famous person he knew whom fame had not corrupted.

She represented a remarkable mixture of curiosity and perseverance. Once she identified a question, she was able to work tirelessly to find an answer, or at least to frame it more fruitfully. To appreciate the strength of her determination, we need only call to mind the image of the diminutive Marie labouring day after day, week after week over reductions, washings, and distillations, all to extract just fractions of a gram of radium from tons of pitchblende.

Above: A tribute to the memory of Marie Curie
– a specially bound edition of the biography
Madame Curie is presented by Marjorie Illig of the
American Society for the Control of Cancer to
Jule Henry, Minister Plentipotentiary of France,
while Eleanor Roosevelt, wife of the US president,
looks on.

What made such dedication so inspiring is not just her refusal to give up but her sense of fulfilment and even exhilaration in the work itself. It was not that Pierre's dedication fulfilled Marie's dreams or that Marie's dreams fulfilled Pierre's. Rather, they participated in the same dream, one so engrossing that they could think of little else. Of the early days of their work together she wrote:

We were very happy in spite of the difficult conditions under which we worked. We passed our days at the laboratory, often eating a simple student's lunch there. A great tranquillity reigned in our poor, shabby hangar; occasionally, while observing an operation, we would walk up and down talking of our work, present and future. When we were cold, a cup of hot tea, sipped beside the stove, cheered us. We lived in a preoccupation as complete as that of a dream.

In Marie's words we hear and through her eyes we see the beauty of a life dedicated to revealing the mysteries of creation. Beyond the comprehensibility of the rules by which nature operates and the ingenuity required to solve experimental challenges lies an elegance that can only be described in aesthetic terms. Science doesn't just tell us how things work – it also opens our eyes to great beauty all around us. It inspires us in a quest to explore and to marvel.

Said Marie in one of her last public addresses in 1933,

I believe that science has great beauty. A scientist in the laboratory is not a mere technician; he is also a child confronting natural phenomena that impress him as though they were fairy tales. We should not allow it to be believed that all scientific progress can be reduced to mechanisms, machines, and gearings, even though such machinery also has beauty. Neither do I believe that the spirt of adventure is in danger of disappearing from our world. If I see anything vital around me, it is this very spirit of adventure, which seems ineradicable and is very closely related to curiosity.

Below: The periodic table of the elements, including curium, an element with atomic number 96, which was named after the Curies.

Periodic Table of the Elements

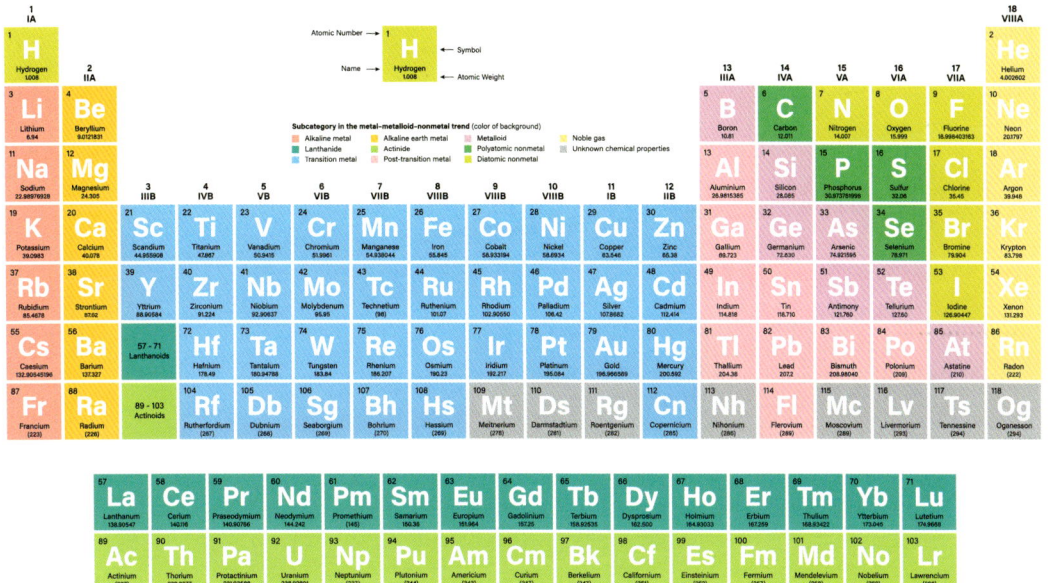

INDEX

(page numbers in **bold** refer to substantial information; *italics* refer to photographs, illustrations and captions)

CREDITS

The publishers would like to thank the following sources for their kind permission to reproduce the pictures in this book.

Key: t = top, b = bottom, l = left & r = right

© AAAS: 138

Académie des Sciences: 93t

Alamy: Chronicle 59, 117; /Ian Dagnall Computing 126t; /GL Archive 4-5; /Granger Historical Picture Archive 63, 72l; /History & Art Collection 75, 81, 92; /The History Collection 81, 127; /Historic Collection 35, 120; /John Frost NewspapersOxford Science Archive/Print Collector 98; /Pictorial Press 78; /The Picture Art Collection 33, 65b, 111; /RGB Ventures/SuperStock 136; /Science History Images 48, 56; /Marcin S. Sadurski 152; /Steve Speller 188; /Björn Wylezich 55

Bibliothèque Nationale de France: 80, 83

Bridgeman Images: 110; /Archives Charmet 32; /Photo © Christie's Images 112; /Giancarlo Costa 37; /Look & Learn 121; /Private Collection 86, 94, 96

Getty Images: Bernard Bailly/AFP 123; /Bain News Service/Buyenlarge 135; /Bettmann 66, 143b, 148; /Buyenlarge 87; /Corbis 126; /Couprie/Hulton Archive 101; /Culture Club 17; /E+ 13; /DeAgostini 154t; /Fine Art Images/Heritage Images 137; /Albert Harlingue/Roger Viollet 52, 102; /Horst P. Horst/Condé Nast 153; /Hulton Archive 29; /Imago 144; /iStock 12; /Keystone-France/Gamma-Rapho 22, 95; /Mondadori 28b, 104; /ND/Roger Viollet Abraham 23; /Oxford Science Archive/Print Collector 84, 147; /Pisarek/ullstein bild 14-15; /Popperfoto 19, 20, 85, 131; /SSPL 49, 67, 105, 106, 107, 109; /Ullstein Bild 46l, 124; /Universal History Archive 8, 31, 128, 139t, 139b

Heritage Auctions, HA.com: 132

Library of Congress: 10, 91, 113, 114-115, 154b

Mauswiesel via Wikimedia Commons 57

Digitized by the Mütter Museum of The College of Physicians of Philadelphia: 54

Photo de famille_Paul-Gilbert_Langevin: 119

NIST (National Institute of Standards and Technology): 141

Private Collection: 18, 21, 26, 27, 68, 88, 93b, 103, 125

Public Domain: 9, 39, 45, 46r, 47, 65, 65t, 72r, 73, 74, 77, 97, 108, 116, 122, 130, 134, 143t

Science Photo Library: 41; /A.Barrington Brown, © Gonville & Caius College 42; /Humanties & Social Sciences Library/New York Public Library 69

Shutterstock: 7, 38, 40; /Roman Belogorodov 150; /Everett Historical 40; /Granger 11, 16; /Humdan 156; /TTStudio 60-61; /Universal History Archive/UIG 79; /Yoan Valat/EPA-EFE 70

Smithsonian Institution @ Flickr Commons: 90, 142

Société Française de Physique: 145

Topfoto: Roger-Viollet 99

U.S. Department of Energy: 149

Every effort has been made to acknowledge correctly and contact the source and/or copyright holder of each picture and Welbeck Publishing apologises for any unintentional errors or omissions, which will be corrected in future editions of this book.

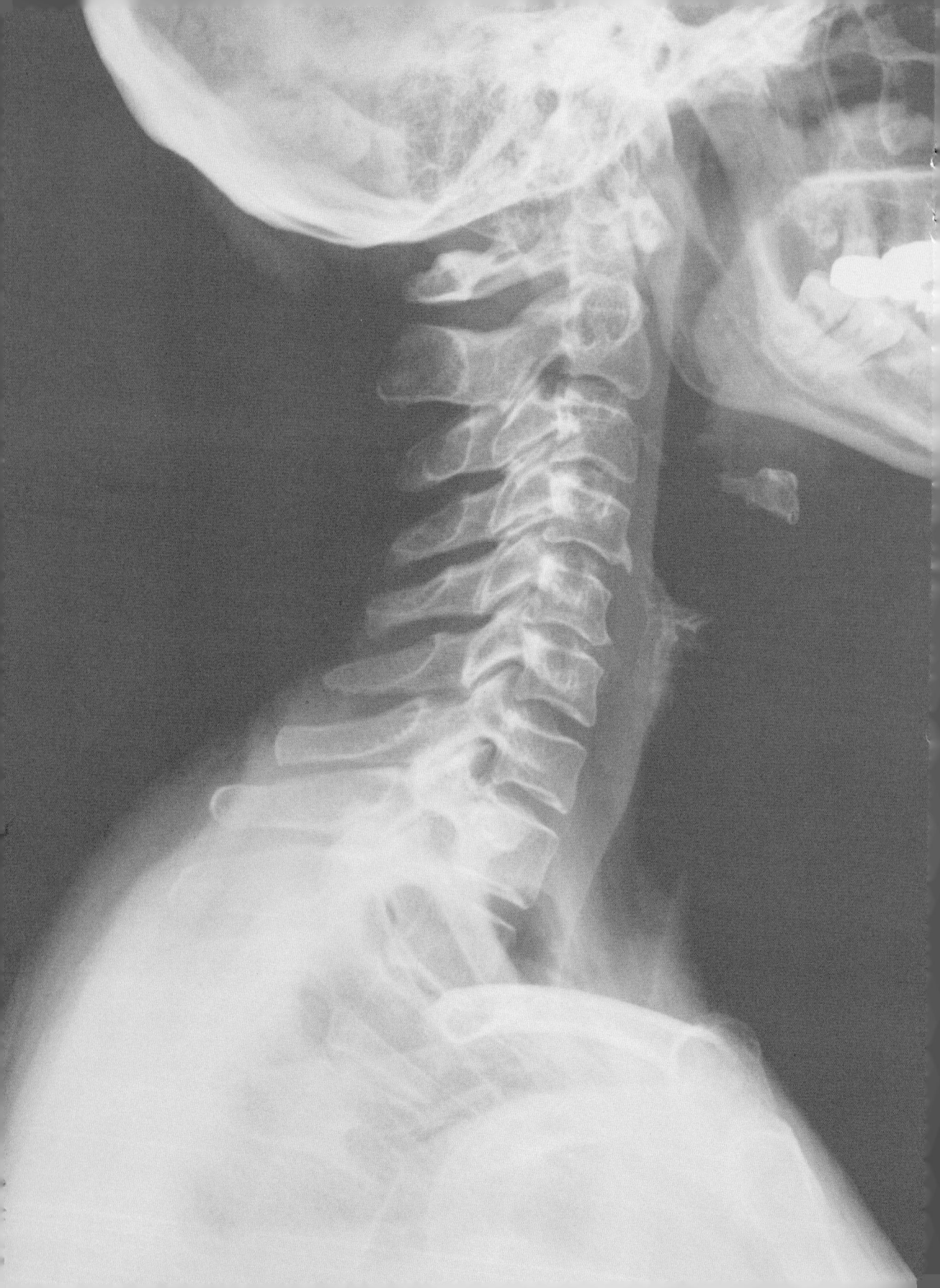